Veröffentlichungen des
Instituts Wiener Kreis

Band 14

Hrsg. Friedrich Stadler

Friedrich Stadler
Kurt R. Fischer (Hrsg.)

Paul Feyerabend

Ein Philosoph aus Wien

SpringerWienNewYork

Ao. Univ.-Prof. Dr. Friedrich Stadler
Universität Wien und Institut Wiener Kreis, Wien, Österreich

Hon.-Prof. Dr. Kurt R. Fischer
Universität Wien, Wien, Österreich

Gedruckt mit Unterstützung des Bundesministeriums für Bildung, Wissenschaft und Kultur in Wien sowie des Magistrats der Stadt Wien, MA 7 – Gruppe Kultur, Wissenschafts- und Forschungsförderung

Das Werk ist urheberrechtlich geschützt.
Die dadurch begründeten Rechte, insbesondere die der Übersetzung, des Nachdruckes, der Entnahme von Abbildungen, der Funksendung, der Wiedergabe auf fotomechanischem oder ähnlichem Wege und der Speicherung in Datenverarbeitungsanlagen, bleiben, auch bei nur auszugsweiser Verwertung, vorbehalten.

© 2006 Springer-Verlag/Wien Printed in Austria
SpringerWienNewYork ist ein Unternehmen von
Springer Science + Business Media
springer.at

Die Wiedergabe von Gebrauchsnamen, Handelsnamen, Warenbezeichnungen usw. in diesem Buch berechtigt auch ohne besondere Kennzeichnung nicht zu der Annahme, dass solche Namen im Sinne der Warenzeichen- und Markenschutz-Gesetzgebung als frei zu betrachten wären und daher von jedermann benutzt werden dürften. Produkthaftung: Sämtliche Angaben in diesem Buch erfolgen trotz sorgfältiger Bearbeitung und Kontrolle ohne Gewähr.

Satz: Reproduktionsfertige Vorlage der Herausgeber
Umschlaggestaltung unter Verwendung einer Fotografie von Paul Feyerabend, aufgenommen um 1947, mit freundlicher Genehmigung von Grazia Borrini-Feyerabend
Druck: Börsedruck Ges.m.b.H., 1230 Wien, Österreich
Gedruckt auf säurefreiem, chlorfrei gebleichtem Papier – TCF
SPIN: 11572565

Mit 2 Abbildungen

Bibliografische Information der Deutschen Nationalbibliothek
Die Deutsche Nationalbibliothek verzeichnet diese Publikation in der Deutschen Nationalbibliografie; detaillierte bibliografische Daten sind im Internet über http://dnb.d-nb.de abrufbar.

ISBN-10 3-211-29759-6 SpringerWienNewYork
ISBN-13 978-3-211-29759-9 SpringerWienNewYork

INHALTSVERZEICHNIS / TABLE OF CONTENTS

Editorial vii

Friedrich Stadler
Paul Feyerabend – Ein Philosoph aus Wien ix

Kurt Rudolf Fischer
Feyerabends Weltanschauung 1

Paul Hoyningen-Huene / Eric Oberheim
Neues zu Feyerabend 13

Erhard Oeser
Paul Feyerabend zwischen Wissenschaftsgeschichte
und Wissenschaftstheorie 35

Reinhold Knoll
Warum Wissenschaft eine Kunst ist ...
Gedanken zu Paul Feyerabend 49

Hans Sluga
Der erkenntnistheoretische Anarchismus.
Paul Feyerabend in Berkeley 61

Karl Svozil
Feyerabend and Physics 75

Juliet Floyd
Homage to Vienna: Feyerabend on Wittgenstein
(and Austin and Quine) 99

Namenregister / Index of Names 153

Die Autoren / The Authors 157

EDITORIAL

Die Beiträge zum vorliegenden Band Paul *Feyerabend – Ein Philosoph aus Wien* gehen zurück auf ein internationales Symposium, das – auf Anregung von Rudolf Haller und Kurt Fischer – vom Institut Wiener Kreis gemeinsam mit dem Institut für Zeitgeschichte und dem Institut für Philosophie der Universität Wien aus Anlass von Feyerabends 10. Todestag am Universitätscampus in Wien veranstaltet wurde.

Ziel dieser Veranstaltung war es, eine kritische Neubewertung seines Lebens und Werkes jenseits eines trendigen Geniekultes auf der Grundlage neuer Quellen und aktueller Forschung zu liefern. Es stellte einen erstmaligen Versuch dar, den viel diskutierten und schillernden Denker vor allem in das Wiener philosophische Milieu der Nachkriegszeit zu stellen und seine fragmentarische Autobiografie in die aktuelle Forschung einzubetten.

Dementsprechend finden wir Beiträge von Feyerabends persönlichen Freunden und Kollegen (Kurt R. Fischer und Hans Sluga) über dessen „Weltanschauung" bzw. die Grenzen seines „erkenntnistheoretischen Anarchismus", von Wien bis zu seiner Zeit in Berkeley und Zürich vor. Daneben eine Studie über Feyerabends Verhältnis zu Wittgenstein, seine vielleicht wichtigste philosophische Bezugsperson (Juliet Floyd); eine Studie, die Feyerabends komplexe Beziehung zu Kuhn am Beispiel des Inkommensurabilitätsbegriffes aus neuer Perspektive thematisiert und dessen Methode und Inhalte einheitlich betrachtet. (Paul Hoyningen-Huene und Eric Oberheim). Weiters wird Feyerabends Bild von „Wissenschaft als Kunst" auf dessen eigenes Denken und Handeln mit Bezügen zur Wiener Wissenschaftskultur angewandt (Reinhold Knoll); seine Präferenz für eine empirische Wissenschaftsgeschichte gegenüber einer normativen Wissenschaftstheorie mit rationalem „Methodenzwang" (Erhard Oeser); schließlich über seine nicht immer klare Position zur Physik des 20. Jahrhunderts im Verhältnis zur Wissenschaftsphilosophie (Karl Svozil).

Insgesamt liefert der vorliegende Band ein neues und differenziertes Bild des „Philosophen aus Wien", der bis zu seinem Lebensende auch ein „Wiener Philosoph" (in der Tradition von Mach bis zu den drei Wiener Kreisen) geblieben ist – und der ohne die prägende erste Dekade nach 1945 zwischen Wien und Alpbach kaum verständlich wird (Friedrich Stadler). Dadurch wurde der kontroversielle Philosoph der Postmoderne zwischen Relativismus und Pluralismus historisiert und zugleich mit neuen Perspektiven auf sein Leben und Werk entmythologisiert.

Die Herausgeber danken Grazia Borrini-Feyerabend für ihre Unterstützung des Symposiums und die Abdruckrechte sowie den Archiven des Österreichischen College in Wien und an der Universität Konstanz mit dem Feyerabend-Nachlass für die Kooperation und Benützung. Außerdem den Mitarbeitern am Institut Wiener Kreis, vor allem Christoph Limbeck-Lilienau für die Literatur- und Archivarbeit und Robert Kaller für die Redaktion und Herstellung des Manuskriptes.

Nicht zuletzt gebührt unser Dank dem Österreichischen Bundesministerium für Bildung, Wissenschaft und Forschung und der Wissenschaftsabteilung der Stadt Wien für die finanzielle Unterstützung der Publikation.

Für die Veranstalter und Herausgeber,

Wien, im August 2006
Friedrich Stadler
(Universität Wien und
Institut Wiener Kreis)

FRIEDRICH STADLER

PAUL FEYERABEND – EIN PHILOSOPH AUS WIEN*

Paul Feyerabend ist – auch 12 Jahre nach seinem Tod – noch immer im Gespräch, in der philosophischen Szene, in der Scientific Community und in der breiteren öffentlichen Diskussion. Besonders in der deutschsprachigen Szene wird sein Image als *enfant terrible* der Philosophie durch posthume fragmentarische Publikationen noch weiter genährt (Feyerabend 2005). Sein Image bleibt komplex und widersprüchlich, als Ikone von „Anti-Science" hat sich sein Lebenswerk verselbständigt und sein Name ist zu einem beliebten polarisierenden Instrument im zeitgeistigen Kulturkampf geworden.

Während die Darstellung und Erforschung seiner Arbeit seit dem Erscheinen seines erfolgreichen Buches *Against Method* (Feyerabend 1975) noch im vollen Gange ist, scheint die Zeit davor – mit wenigen Ausnahmen (z.B. Haller 1997; Hochkeppel 2006) überraschenderweise noch wenig behandelt. Das ist umso bemerkenswerter, als gerade diese Phase prägend für die intellektuelle Entwicklung gewesen ist, und man kann sogar behaupten, dass eine Rückkehr des späten Feyerabend zu ursprünglichen Themen seiner Wiener Zeit zu erkennen ist. Kurz gesagt: es gibt mehr Kontinuität als Bruch im Biografischen und mehr Konsistenz als Widerspruch im Theoretischen in Feyerabends philosophischem Werk.

Meine Hauptthese gegen den üblichen Forschungstrend in Richtung poststrukturalistische Interpretation (wie z.b. ein eigener Workshop 30 Jahre nach *Against Method*: Grebovicz 2006) ist also, dass Feyerabend Zeit seines Lebens trotz erheblicher intellektueller Wandlungen eigentlich ein „Philosoph aus Wien" oder „Wiener Philosoph" geblieben ist, auch wenn er vielleicht diese Aussage so nicht unterschrieben hätte. Darüber hinaus ist er, und das ist meine Zusatzthese, auch nach seinem Weggang aus Wien der österreichischen Philosophie und mitteleuropäischen Wissenschaftstradition entscheidend verhaftet geblieben, sodass seine geistige Entwicklung ohne diesen Kontext schwer verständlich wird. Das könnte man bereits aus seiner fragmentarischen Autobiografie *Killing Time* (1994, deutsch: *Zeitverschwendung* 1995) erschließen, die jedoch aufgrund der Gewichtung und der identitätsstiftenden Diktion – wie jede Selbstdarstellung – als ausschließliche Quelle problematisch ist und sehr oft in die klischeehafte Irre führt. Unbestritten sind dort die wesentlichen Themen und Probleme seines Lebensweges präsent, auch wenn das impressionistische Arrangement der Erinnerungsstücke in ein teleologisches Narrativ des Ich-Erzählers einmündet. Dies soll im folgenden kurz aus biografischer und philosophischer Sicht illustriert werden

Feyerabend in Wien: Der vergessene „Dritte Wiener Kreis" 1948– 1954

Nach einem kurzen Aufenthalt mit künstlerischen Ambitionen in Deutschland (Weimar) kehrte Feyerabend nach Wien zurück und begann sein Studium 1947 an der Universität Wien: zuerst inskribierte er Geschichte und Soziologie, kurz darauf Physik bei Hans Thirring, Karl Przibram und Felix Ehrenhaft und Philosophie bei Alois Dempf und besonders Viktor Kraft. (Dazu autobiografisch Feyerabend 1995, 87-110).

In diesem Zusammenhang bot sich für die intellektuelle Entwicklung des jungen Studenten eine wichtige institutionelle internationale Plattform an, das von den Brüdern Fritz und Otto Molden 1945 gegründete und bis heute existierende „Österreichische College / Forum Alpbach", welches sowohl in Wien als auch im Tiroler Bergdorf Alpbach eine rege Veranstaltungstätigkeit entwickelte. (Molden 1981; Auer 1994) Dort kam Feyerabend erstmals ab 1948 in Kontakt mit den großteils exilierten Wissenschaftsphilosophen sowie mit der übrigen scientific community, unter ihnen Karl Popper, Friedrich von Hayek, Hans Albert, und von den früheren Wiener Kreis-Mitgliedern Rudolf Carnap, Herbert Feigl, Philipp Frank. Auch für den in Österreich verhinderten Wolfgang Stegmüller bildete Alpbach den geistigen Einstieg in das internationale Netzwerk der wissenschaftlichen Philosophie und Wissenschaftstheorie.

Im Rahmen dieses einzigartigen Forums zur wissenschaftlichen und kulturellen „Entprovinzialisierung" am Beginn der Zweiten Republik wurde ein eigener „Arbeitskreis für Naturphilosophie" gegründet, der als „Kraft-Kreis" von 1949–1953 eine regelmäßige Diskussions- und Publikationstätigkeit in Wien entfachte. Viktor Kraft war der akademische Leiter und Feyerabend fungierte als Sprecher der Studierenden. Aufgrund der Teilnehmer aus dem In- und Ausland und der Wirkungsgeschichte ist es keine Übertreibung, von einem kurzfristigen bislang kaum wahrgenommenen „Dritten Wiener Kreis" zu sprechen. (Stadler 2006): so finden wir als Mitglieder noch Bela Juhos, Walter Hollitscher, Ernst Topitsch neben den Studenten J. Sagan, H. Eichhorn, Goldberger de Buda, P. Schiske, Erich Jantsch, sowie unter den Gästen u.a. Elisabeth Anscombe, Emil J. Walter, Georg Henrik von Wright, E. Tranekjaer-Rasmussen und zumindest einmal noch Ludwig Wittgenstein (wahrscheinlich 1950). In seinen Erinnerungen schreibt Feyerabend dazu:

> Kraft war ein führendes Mitglied des Wiener Kreises. Wie Thirring wurde er nach dem „Anschluss" Österreichs in den Ruhestand versetzt. Er war ein mäßiger Redner, aber ein kluger und sorgfältiger Denker. Er hat einige Ideen vorweggenommen, die später Popper zugeschrieben wurden ...
> Er kannte die meisten von uns aus dem Seminar und drückte den Wunsch nach regelmäßigen Treffen aus. So kam es zur Bildung des Kraft-Kreises, einem studentischen Pendant des alten Wiener Kreises. Wir erhielten einen Raum in der Kolingasse, dem Büro des Österreichischen College, und trafen uns zweimal im Monat. Wir diskutierten über konkrete wissenschaftliche Theorien. Zum Beispiel behandelten wir in fünf Sitzungen die nicht-Einsteinschen Interpretationen der Lorentztransformationen. Unser Hauptthema war die Frage der Existenz einer Außenwelt. (Feyerabend 1995, 104).

Rückblickend kritisiert Feyerabend die inhaltliche Auffassung von Wissenschaft als System von Sätzen auf der Basis seiner Lektüre der Zeitschrift *Erkenntnis*. Im wesentlichen ging es um die Rechtfertigung eines kritischen/konstruktiven Realismus mit hypothetisch-deduktiver Methodologie, die Kraft bereits in seiner Schrift *Die Grundformen der wissenschaftlichen Methoden* (1925) behandelt hatte. (Über Kraft im Vergleich zu Feyerabend: Radler 2006).

All diese Diskussionen spiegeln sich sehr deutlich in der (unveröffentlichten) Dissertation *Zur Theorie der Basissätze* (1951), die von Kraft betreut wurde. In seinem dem Rigorosenakt beigelegten Lebenslauf wird die persönliche Motivation Feyerabends sowie die problemgeschichtliche Ausgangslage angesprochen, sodass dieser hier nachfolgend wörtlich wiedergegeben sei:

Ich, Paul Feyerabend, wurde am 13.1.1924 in Wien geboren, besuchte die Volks- und Mittelschule. Mein Interesse an Philosophie folgte einem lebhaften Interesse für Naturwissenschaften. So gehörten zu meiner Lektüre Duhem, Mach und Dingler.

1942 bis 1945 Militärdienst, hierauf ein Jahr Lazarett, ein Jahr Studium an der staatlichen Musikhochschule Weimar, dann Rückkehr nach

Wien, 1 Semester Geschichte und Kunstgeschichte, 6 Semester Astronomie, Physik und Mathematik, schliesslich Übertritt zur Philosophie. Hierbei waren mir vor allem die Diskussionen, die in einem kleinen Kreis um Prof. Kraft stattfanden und in denen nach Art des Wiener Kreises Probleme der Wissenschaftstheorie vor allem behandelt wurden, von grossem Nutzen. Hier und durch Anregung, die ich von Prof. Popper (London School of Economics) erhielt, begann mich das in der Dissertation bearbeitete Thema zu interessieren. Ich hatte seit 1948 öfter Gelegenheit an Diskussionen teilzunehmen, aus denen ich viel für die endgültige Fassung dieser Arbeit gewinnen konnte. Vor allem waren mir da die Unterredungen mit Prof. Walter Hollitscher (Berlin) von grossem Nutzen. Wenn nicht durch seine Argumente, so kam ich doch durch seinen wiederholten Ansturm dazu, meine philosophischen Ansichten genauer zu überprüfen und (vom Positivismus Mach'scher Art bis zu dem hier vertretenen Standpunkt) wesentlich zu korrigieren. Desgleichen warn mir die Diskussionen mit Mrs Anscombe (Cambridge) über die Problematik des UG von grossem Nutzen. Damals präsentierte sie mir zahlreiche Wendungen, die mir völlig unverständlich erschienen und die ich laengere Zeit unverdaut mit mir schleppte (ebenso wie einige Wendungen, die ich aus einer Diskussion mit L. Wittgenstein, der an einem Abend an den Diskussionen um Prof. Kraft teilnahm, mitnahm).

Im Lauf der Zeit, weniger durch Nachdenken, als durch einen unbewussten Entwicklungsprozess fanden sich Möglichkeiten des Verständnisses. Sie sind in der Dissertation dargestellt. Sie erscheinen mir gegenwärtig als die richtige Interpretation jener Wendungen (was nicht die historische Richtigkeit jener Interpretation zur Folge haben braucht). Die Grundgedanken einer früheren Fassung konnte ich im Rahmen eines Vortrags in der philosophischen Gesellschaft Uppsala und in einem kleinen Kreis in Kopenhagen mit Prof. Marc-Wogau und Prof. Joergensen (mit letztem auch privat) diskutieren. Beiden Diskussionen habe ich viel zu verdanken.

Prof. Kraft hat mich auf schauderhafte Verworrenheiten einer früheren Fassung sowie auf mehrere Unklarheiten aufmerksam gemacht.

Für wesentliche Züge der Grundeinstellung bin ich Prof. Tranekjaer-Rasmussen (Kopenhagen) sehr zu Dank verpflichtet. Er erlaubte mir zwei noch unveröffentlichte Manuskripte zu lesen, die ausgeführt enthalten, was er auf einem Vortrag in Alpbach 1948 (auf den auch mehrfach verwiesen wird) kurz angedeutet hat.

Ich hoffe, dass es mir bald möglich sein wird, auf der Basis dieser (noch unvollkommenen) Vorarbeit bald zunaechst eine Theorie physikalischen Wissens aufzufinden

(Gez. Paul Feyerabend)

Diese bislang kaum wahrgenommene frühe prägende Arbeit von Feyerabend fand laut Rigorosenakt Nr. 18.107 der Universität Wien die volle Zustimmung durch den Erstbegutachter Viktor Kraft:

„Die Diss. zeugt von einer außergewöhnlichen Begabung. Diese spricht schon aus dem von der Schablone gänzlich abweichenden Lebenslauf. Das Thema der Arbeit bildet die Rolle der Wahrnehmungsaussagen bei der Verifikation in den empirischen Wissenschaften, die durch den modernen Empirismus zur Diskussion gestellt worden ist. Die Untersuchungen der Diss. greifen aber über diese fundamentale Aufgabenstellung noch weit hinaus, indem sie von da aus die Bestimmung der Gegenstände der Physik und der nicht-physikalischen Wissenschaften, bes. der Psychologie, aufnehmen und damit den Gegensatz von Phänomenalismus und Physikalismus zur Lösung bringen. Es ist der Grundgesichtspunkt der Arbeit, zweierlei Aspekte der Wahrnehmungsaussagen klar zu sondern: einerseits die Charakterisierung der Wahrnehmungsaussagen als ausgezeichnet durch einen unmittelbar gegebenen Kern, anderseits ihre verifizierende Funktion. In der ersten Hinsicht wird die Rolle der Wahrnehmung dain bestimmt, dass sie eine bestimmte Aussage veranlasst, sie erschöpft sich in einer bloßen Motivationsfunktion dafür. Durch sie wird der Aussage-Inhalt aus dem Erleben verständlich. Aber eine Wahrnehmungsaussage wird dadurch nicht logisch vor anderen ausgezeichnet, sie erhält dadurch keine zweifelsfreie Geltung, wie es die Theorie der „Konstatierung" annimmt. Sie muss wie jede andere wissenschaftliche Aussage überprüft werden. Das Prüfungsverfahren wird eingehend analysiert und eine eigene Theorie der verlässlichen Beobachtung entwickelt. Damit wendet sich der Verf. in ausführlicher Kritik gegen die überwiegend vertretene Anschauung, dass die Wahrnehmungsaussage das logische Fundament der empirischen Erkenntnis bildet. Er weist die unentbehrliche Voraussetzung auf, dass immer eine <u>Theorie</u> die Grundlage für die Verwertung einer Wahrnehmungsaussage herstellt; nur innerhalb einer Theorie erhält diese eine bestimmte logische Funktion. Im Grundsätzlichen sind die Ergebnisse der Arbeit durchaus anzuerkennen: sie sind neu und haben wirklichen, bleibenden Wert. Die Ausführungen bewegen sich auf einem ganz außergewöhnlich hohen Niveau, sie werden mit gro-

ßem logischen Scharfsinn, mehrfach, besonders im Schlussteil, in logistischer Entwicklung, geführt; der Verf. kennt die relevante moderne angelsächsische und skandinavische Literatur ausgezeichnet, zieht aber auch klassische Philosophen in Originalstellen heran. Bei der Fülle der erörterten Fragen ist die Darstellung sehr gedrängt ausgefallen, und oft voraussetzungsvoll, so dass sie deshalb nicht immer leicht verständlich ist. Die Arbeit erhebt sich weit über den Rang einer gewöhnlichen Diss. hinaus und muss als ganz ausgezeichnet bewertet werden."
Wien, 12. Juni 1951.

(Gez. V. Kraft, F. Kainz)

Das Rigorosum in den Fächern Philosophie und Psychologie mit den Prüfern Viktor Kraft, Friedrich Kainz und Hubert Rohracher ergab insgesamt ein „ausgezeichnet", die einstündige Prüfung in der Physik insgesamt ebenfalls ein „ausgezeichnet" von Hans Thirring und E. Schmid.

Was hier theoriegeschichtlich auffällt, ist die originelle Rekonstruktion des Basisproblems samt Protokollsatzdebatte im Wiener Kreis unter Einbeziehung zeitgenössischer experimentalpsychologischer Forschung mit einer pointierten Reformulierung der Duhem-Neurath-Quine These von der prinzipiellen Theoretizität aller empirischer Aussagen, also einer aktuellen Diskussion der Quine'schen „Two Dogmas of Empiricism" (1951). Tatsächlich findet sich im Archiv des Österreichischen College ein entsprechendes Manuskript Feyerabends mit dem Titel „Die Dogmen des Logischen Empirismus", das die Grundlage eines Vortrags im gleichen Jahr bildete.

In den Jahren 1949–52 unternahm Feyerabend seine ersten Auslandsreisen nach Skandinavien, wo er sich mit Louis Hjelmslev, Tranekjaer-Rasmussen, Joergen Joergensen, Marc-Wogau, Wedberg u.a. traf. In diesem Zusammenhang ist zu bemerken, dass die seit der Jahrhundertwende um 1900 gepflegte intensive Wissenschaftskommunikation zwischen Mitteleuropa und den nordischen Staaten – im Gegensatz zur angelsächsischen Welt – eine beachtliche, wenig beschriebene wechselseitige Kontinuität vor, während und nach dem Zweiten Weltkrieg aufweist und insofern eine Ausnahme im Zusammenhang mit dem „Cultural Exodus" des Logischen Empirismus darstellt. Allein die bedeutende Rolle von Eino Kaila und Arne Naess ist ein Indiz für diese in der Historiografie marginalisierte „Nordic Connection", die sich auch in der von Ake Petzäll herausgegebenen Zeitschrift *Theoria* (1935ff.) manifestiert und heute vor allem durch Jaakko Hintikka und seine Schüler fortgesetzt wird. Es ist also kein Zufall, dass Kaila mehrmals den Schlick-Zirkel besuchte und in seinen Schriften im Sinne eines erkenntnistheoretischen Realismus auch übereinstimmend mit Kraft Kritik am so genannten „Logistischen Neupositivismus" (1930) übte. Desgleichen erscheint die frühe Auseinandersetzung von Naess mit Neurath und Carnap im Rahmen seiner Dissertation und in *Erkenntnis und wissenschaftliches Verhalten* (1936) wie eine Vorwegnahme von Feyerabends Skepsis gegenüber Wissenschaft als Aussagensystem, auch wenn dieser von Naess, dem späteren Philosophen der Öko-Bewegung, gerade nicht als ein „Held der Grünen" anerkannt werden sollte. (Naess 1980). Naess war es übrigens auch, der vergeblich nach dem Ausbruch des Zweiten Weltkrieges den sechsten und letzten „International Congress for the Unity of Science" in Oslo veranstalten wollte, nach dem der zweite Kongress in Kopenhagen unter der Ägide von Niels Bohr und Joergensen im Jahr 1936 stattgefunden hatte. Jedenfalls waren es die Themen der nordischen Philosophen wie Realismus/Materialismus und das Verhältnis von Psychologie und Philosophie, welche im klassischen Wiener Kreis des Schlick-Zirkels und im Dritten Wiener Kreis eine wesentliche Rolle spielen, und die auch Feyerabend mehr oder weniger beschäftigen sollten.

Im Studienjahr 1953/54 kam auf Initiative des bereits emeritierten Kraft noch Arthur Pap (1921–1959) als Fulbright Gastprofessor an das Institut für Philosophie der Universität Wien und ermöglichte eine kurzfristige Renaissance der analytischen Philosophie in der Tradition des vertriebenen Logischen Empirismus. (Pap 1955). Als Assistenten engagierte der viel versprechende Philosoph den begabten Paul Feyerabend, der dessen Vorlesungsmanuskripte für das Buch *Analytische Erkenntnistheorie. Kritische Übersicht über die neueste Entwicklung in den USA und England* 1955 im Springer Verlag redigierte. Dieses, „dem ‚Wiener Kreis' zum Andenken und zur Wiederbelebung" gewidmete Werk stellte eine Rekonstruktion und kritische Weiterentwicklung der wissenschaftlichen und sprachanalytischen Philosophie nach dem Exodus des Logischen Empirismus dar – und zwar ebenfalls von einem Emigranten, dessen Familie nach Ausbruch des Zweiten Weltkriegs wegen ihrer jüdischen Herkunft und aus politischen Gründen aus der Schweiz (Zürich) in die USA ausgewandert war.

Dort studierte Pap an der Columbia University und in Yale u.a. unter Ernst Cassirer, Ernest Nagel und Charles Stevenson. Später kam er in Kontakt mit Rudolf Carnap und Herbert Feigl, mit denen er eine lebenslange Freundschaft pflegte. Zuvor hatte er auch Krafts *Der Wiener Kreis* ins Englische übersetzt (1953) und absolvierte nach seiner Rückkehr 1955 in die USA einen Forschungsaufenthalt an Feigls „Minnesota Center for Philosophy of Science", bevor er in Yale Nachfolger von Carl G. Hempel wurde. Sein letztes Buch konnte wegen seines allzu frühen Todes nur mehr postum unter dem Titel *An Introduction to the Philosophy of Science* (1962) erscheinen.

Die Hoffnung von Arthur Pap, der damals schon in den USA trotz seiner Jugend ein renommierter Vertreter der analytischen Philosophie gewesen ist, auf ein permanentes akademisches Engagement an der Universität Wien hat sich nicht erfüllt und er verstarb aufgrund einer heimtückischen Nieren-Krankheit sehr jung im Alter von 38 Jahren. (Keupink/Shie 2006).

Für Wien hätte sein Verbleiben sicherlich eine länger anhaltende Wirkung des „Dritten Wiener Kreises" mit einem möglichen Engagement von Feyerabend bedeutet. Letzterer war nämlich – er wollte ursprünglich bei Wittgenstein studieren – 1952 an der LSE bei Karl Popper gewesen, der ihn als Assistenten beschäftigen wollte. Feyerabend lehnte ab, aber er übersetzte Poppers umfangreiches zweibändiges Werk *Open Society* ins Deutsche. Eine der ersten Besprechungen von Wittgensteins *Philosophische Untersuchungen* im Kontext der österreichischen Philosophie (Ernst Mach, Heinrich Gomperz) ist ein Resultat dieser Aktivitäten. (Feyerabend 1955). Die erste akademische Position in Bristol (UK) im Jahre 1955 bedeutete den Beginn von Feyerabends internationaler Karriere und de facto das Ende der kurzfristigen Renaissance des Wiener Kreises in der Zweiten Republik, wiewohl die Kontakte zu Wien und Alpbach nie abbrechen sollten. Vor seinem Weggang schrieb er noch ein umfangreiches, informatives und ausgewogenes Manuskript über die „Geisteswissenschaften in Österreich" (1954), in dem er im Kapitel über die Philosophie sehr differenziert die Eigenheiten und Defizite der Nachkriegszeit beschreibt.

Im gleichen Jahr konnte Feyerabend in Wien durch Paps Vermittlung noch Herbert Feigl kennen lernen, was für seine weitere Karriere und den „Wiener Kreis in America" (Feigl 1968/69) im Minnesota Center sehr prägend werden sollte. In der von ihm mit Grover Maxwell herausgegebenen Festschrift für Feigl schreibt er:

> I first met Herbert Feigl in 1954, in the pleasant and stimulating atmosphere of a Vienna coffeehouse. I was then an assistant to Arthur Pap, who had come to Vienna to lecture on analytic philosophy and who had hoped, perhaps somewhat unrealistically, that he would be able to revive what was left from the great years of the Vienna Circle and the analytic tradition there ... After a lecture, which

frequently turned into a heated debate with the attending metaphysicians, we would both retire to the Professorenzimmer of the department of philosophy and discuss what just happened. Pap was alternately depressed and incensed what he thought was the impertinence of those who approached philosophical problems without any knowledge of logic and of analytical techniques, and he contrasted their easily produced Sprachträumereien with the much more modest results which analytic philosophers had achieved by hard work. (1966, 3).

Eine letzte Manifestation der philosophischen Subkultur des Kraft-Kreises stellte dessen Festschrift dar, die – hrsg. von Ernst Topitsch (1960) – letztmalig Schüler und Anhänger Krafts in einem bemerkenswerten Band über *Probleme der Wissenschaftstheorie* versammelte: neben dem Herausgeber noch Hans Albert, Franz Austeda, Rudolf Freundlich, Bela Juhos, Hubert Schleichert, Wolfgang Stegmüller, Emil J. Walter. Auch Feyerabend war mit einem Beitrag vertreten, der ein zentrales Thema des Kraft-Kreises extensiv behandelte: „Das Problem der Existenz theoretischer Entitäten". Auslöser dieser mehrjährigen Dskussion war Herbert Feigls „Existential Hypothesis: Realistic vs. Phenomenalistic Interpretations" (1950) gewesen, im wesentlichen eine Rechtfertigung des „empirischen Realismus". In seinem Postscript (1981) wird eine Kontinuität des wissenschaftstheoretischen Realismus als eine Variante des Logischen Empirismus seit dem frühen Schlick sichtbar und in die zeitgenössische Philosophy of Science bis zu Wilfrid Sellars gestellt, dessen einschlägige Beiträge er am stärksten favorisiert. Feyerabend, bereits 1958 nach Berkeley berufen, problematisiert nochmals das Verhältnis von theoretischen und Beobachtungsbegriffen unter der Annahme einer Korrespondenztheorie mit Kausaltheorie der Erfahrung: „Das ist also die Lösung, die wir für das Problem der theoretischen Entitäten vorschlagen: Jede Beobachtungssprache enthält theoretische Elemente (das folgt aus der Zurückweisung der Sinnesdaten)." (Feyerabend 1960, 70).

Österreichisches College – Forum Alpbach 1955–65

Die philosophische Parallelaktion zum Kraft-Zirkel stellte das bereits erwähnte „Forum Alpbach" des „Österreichischen College" dar: Dort erkennt man im ersten Jahrzehnt eine innovative Renaissance der vertriebenen und zerstörten Wissenschaft durch den Dialog von österreichischen Philosophen mit ausländischen Gastforschern, meist Emigranten. Somit kann man von einer kurzfristigen intellektuellen Gegenöffentlichkeit sprechen, welche die konservativ-klerikalen „langen fünfziger Jahre" im Kontext des Kalten Kriegs kontrastierte. (Hanisch 1995).

Durch die gleichzeitige Präsenz sowohl der traditionellen Philosophie wie auch liberalistischer (Hayek, Machlup, Haberler) und neomarxistischer (Adorno, Horkheimer, Marcuse, Bloch) Denker kam es zu einer intensiven Wiederbelebung der Emigrantenwissenschaften, u.a. mit neun Nobelpreisträgern. Dabei stellten die Auftritte von Rudolf Carnap (1964), Herbert Feigl (1961, 1964), Philipp Frank (1955) und Karl Popper (ab 1948 regelmässig) Höhepunkte der jährlichen Wissenschaftstreffen dar – mit Erwin Schrödinger als Dauergast inmitten der „Geistigen Provinz" (1955) der Nachkriegszeit. Eine direkte Wirkungsgeschichte ist nicht leicht zu rekonstruieren, zumindest *eine* Manifestation ist die von Simon Moser, zusammen mit Otto Molden, dem Begründer des Forum Alpbach, im Vieweg Verlag herausgegebene Buchreihe „Wissenschaftstheorie, Wissenschaft und Philosophie", in der wichtige Werke z.B. von Hans Reichenbach, Imre Lakatos, Heinz von Foerster, und nicht zuletzt Paul Feyerabends ausgewählte Schriften in zwei Bänden – *Der wissenschaftstheoretische Realismus und die Autorität der Wissenschaften* (1978) und *Probleme des Empirismus* (1981) – erschienen sind. Feyerabend schreibt selbst über die Bedeutung Alpbachs und des Österreichischen College, das er 1948 erstmals besuchte, für seine geistige Sozialisation in seiner Autobiografie (97ff.):

> Die Gesellschaft war 1945 gegründet worden, und zwar auf Initiative von Otto Molden, Fritz Molden ... und anderer Mitglieder des österreichischen Widerstands. In seinem Buch *Der andere Zauberberg* beschreibt Otto Molden die Ideen und die Ereignisse, die zum ersten Treffen in Alpbach, einem kleinen Dorf bei Brixlegg in Tirol, geführt haben. Alpbach wurde bald ein internationales Zentrum des geistigen, künstlerischen und wirtschaftlichen Austauschs. Ein Student, der zu Mittag essen ging, konnte einen Platz neben Lise Meitner, Bruno Kreisky oder Dirac finden. Er konnte Arthur Köstler, Anneliese Maier oder Ernst Krenek in die Arme laufen ... Ich habe Alpbach etwa fünfzehnmal besucht, zuerst als Student, dann als Dozent und schließlich dreimal als Leiter eines Seminars.

Bei seinem ersten Besuch im August 1948 lernte der junge Student bereits Karl Popper kennen, der ihn im Vergleich mit den deutschen

Professoren sehr beeindruckte. Diese frühe Wertschätzung beruhte auch auf Gegenseitigkeit, was das spätere Angebot Poppers an Feyerabend, sein Assistent an der LSE zu werden, dokumentieren sollte.

In Alpbach kam der Denkmalstürmer auch mit den beiden österreichischen Philosophen und Kommunisten Hans Grümm (1992) und Walter Hollitscher (Klahr Gesellschaft 2001) in Kontakt, die ihn letztlich vom Realismus überzeugen sollten. Vor allem mit dem Schlick-Schüler Hollitscher, damals noch Psychoanalytiker und Philosoph, verband ihn bis zuletzt trotz politischer Differenzen eine tiefe intellektuelle Freundschaft. Über den Einfluss des letzteren erinnert sich Feyerabend (1995, 100f.):

> Walter Hollitscher hat zwei Jahre gebraucht, um mich davon zu überzeugen, dass Zirkularität ein Gewinn und eine praktische Bereicherung ist, und nicht etwa ein Nachteil. Walter wies mich darauf hin, dass wissenschaftliche Forschung mit realistischen Begriffen betrieben wird. Ich erwiderte, dass die Wissenschaftler leider ihre metaphysischen Eierschalen noch nicht abgeworfen haben. Unabhängig von der Metaphysik haben die Wissenschaftler Ergebnisse vorzuweisen, die von jedem akzeptiert werden, sagte Walter: das gilt auch für die Positivisten, während eine antiseptische Sprache und strenge Logik uns überhaupt nicht weiterbringen. Das Argument brachte mich eine Zeitlang zum Schweigen, aber ein Rest von Zweifel blieb doch erhalten. Von der Physik ging Walter zur Politik über, und das hieß: Marx und Lenin. In diesem Punkt widerstand ich seinen Reden wie ein sturer Esel.

Durch Hollitscher kam Feyerabend übrigens auch mit Bertolt Brecht in Kontakt, dessen Angebot er ausschlug, sein Assistent in Berlin zu werden – was er im nachhinein als größten Fehler seines Lebens bezeichnete.

In einem Brief an Hollitscher aus dem Jahre 1977 wird Feyerabend sich gegen den Vorwurf des Irrationalismus – vorgebracht von K.-P. Noack – von *Against Method* verteidigen (Feyerabend an Hollitscher, 5.9.1977):

> Alle professionellen Wissenschaftskritiker haben sich auf meine Witze gestützt und diese philosophisch analysiert, die *Anlässe* zu den Witzen, d.h. das historische Material hat niemand besprochen und manche haben sogar gesagt, das gehöre nicht zur Sache (d.h. zur Wissenschafts*theorie* ≠ Wissenschafts*geschichte*). Herr Noack

sieht die Witze für was sie sind und kritisiert die Argumente ... Die
Hauptthese des Buches, die ich allerdings nie ausdrücklich formu-
liere, ist: die Bewertung methodologischer (und logischer) Regeln
ist der konkreten Forschung zu überlassen. Forscher erfinden und
untersuchen materielle wie auch intellektuelle Messinstrumente.
Die abstrakten Überlegungen von Philosophen zieht man natürlich
in Betracht, aber man hält sie nicht schon darum für brauchbar, weil
es philosophische oder logische Argumente für ihre „Rationalität"
gibt. Diese Idee ist ja gar nicht neu, scheinbar aber vielen Wissenschaftstheoretikern unbekannt.

In Alpbach war Feyerabend sehr schnell in den Konferenzbetrieb invol-
viert: im Jahre 1955 kam der Physiker und Wissenschaftsphilosoph
Philipp Frank, ehemaliges prominentes Wiener Kreis-Mitglied und Ein-
stein-Biograf, mitten im Kalten Krieg aus Harvard (Reisch 2004) zur
Leitung der Arbeitsgemeinschaft „Erkenntnis und Handlung" nach Tirol.
Dazu schrieb Feyerabend (1955, 140):

> Philipp Frank war ein Vergnügen. Er war sehr gebildet, intelligent,
> witzig und ein geschickter Erzähler. Wenn er die Wahl hatte, ein
> Problem mit einer Geschichte oder einer analytischen Beweisfüh-
> rung zu erklären, dann entschied er sich immer für die Geschichte.
> Einige Philosophen schätzten das gar nicht. Aber sie übersahen,
> dass die Wissenschaft auch eine Geschichte ist, nicht ein logisches
> Problem. Frank führte aus, dass die aristotelischen Einwände ge-
> gen Kopernikus mit der Empirie übereinstimmten, während Galileis
> Trägheitsgesetzt das nicht tut. Wie in anderen Fällen schlummerte
> diese Bemerkung jahrelang latent in meinem Geist; dann begann
> sie zu wuchern. Die Galilei-Kapitel in *Wider den Methodenzwang*
> sind ein spätes Ergebnis dieser Idee.

Obwohl im Falle von Alpbach – im Vergleich zum Kraft-Kreis – nicht
von einer einheitlichen schulebildenden Wissenschaftsphilosophie gesprochen werden kann, ist die regelmässige Thematisierung von Philosophie, Wissenschaft und Weltanschauung samt deren Wechselwirkung und Grundlagen im Rahmen eines internationalen Forums ohne
Zweifel einer der wichtigsten Faktoren für den Re-Transfer bzw. den
Neubeginn der modernen Wissenschaftstheorie in Österreich nach dem
Zweiten Weltkrieg gewesen:

Besonders wichtig war die Tatsache, dass Vertreter der Philosophie und der Einzelwissenschaften, die in den dreißiger Jahren vor allem wegen der politischen Entwicklung aus Österreich und anderen europäischen Ländern emigriert waren und inzwischen internationale Anerkennung gefunden hatten, nach Alpbach kamen – zum Beispiel die Philosophen Karl Popper, Herbert Feigl, Rudolf Carnap, Philipp Frank, Walter Kaufmann, Karl Löwith, Theodor W. Adorno, Max Horkheimer, Herbert Marcuse und Ernst Bloch, der Jurist und Rechtsphilosoph Hans Kelsen, der Soziologe Theodor Geiger, die Ökonomen Friedrich August von Hayek, Fritz Machlup und Gottfried Haberler. Sie trafen dort jüngere Wissenschaftler wie Wolfgang Stegmüller, Ernst Topitsch, Paul Feyerabend, Bernulf Kanitscheider, Paul Weingartner, Karl Acham und Rudolf Wohlgenannt, die sich für ihre Arbeiten aufgeschlossen zeigten. Und sie konnten darüber hinaus mit Vertretern aus dem Ostblock ... diskutieren, die diese Gelegenheit gerne wahrnahmen. (Albert 1994, 18).

Mit dieser Charakterisierung eines selbst beteiligten Wissenschaftstheoretikers der zweiten Generation wird plausibel, wie die dortige Wissenschaftskultur über eine jüngere Generation österreichischer Philosophen einen mehr oder weniger starken Einfluss ausgeübt hat – was im Falle von Feyerabend und Stegmüller besonders gut demonstriert werden kann. (Stegmüller 1979; dazu auch das Forschungsprojekt http://www.univie.ac.at/ivc/stegmueller).

Feyerabend leitete bereits nach seinem Weggang aus Österreich im Jahr 1956 die Arbeitsgemeinschaft „Stetige und unstetige Veränderungen in der Natur" und zusammen mit Herbert Feigl die Arbeitsgemeinschaft „Grundlagenforschung und Einzelforschung" (1964) sowie ein Jahr darauf allein eine weitere mit dem Titel „Philosophie der exakten Naturwissenschaft".

Vor allem der bereits 1931 nach Amerika emigrierte Schlick-Schüler Herbert Feigl sollte für Feyerabend zu den wichtigsten philosophischen Bezugspersonen werden, den er – wie oben zitiert – im Jahre 1954 über Arthur Pap in Wien kennen lernte. Besonders relevant wurde Feigl nach Feyerabends Engagement in Berkeley als dessen Gastgeber in seiner Rolle als Begründer und Leiter des „Minnesota Center for the Philosophy of Science" in Minneapolis (ab 1953). Resultate seiner beiden längeren Aufenthalte im MCPS (1957, 1959) sind seine Aufsätze „Explanation, Reduction and Empiricism" (1962) und vor allem die lange Zeit ignorierte erste Fassung von „Against Method" (1970) zu nennen. Besonders die biografische Skizze über Feigl in der von ihm zusammen mit Grover Maxwell herausgegebenen Feigl-Festschrift *Mind, Matter and Method. Essays in Philosophy of Science in Honor of Herbert Feigl* (1966) sind aufschlussreiche Dokumente dieser verschütteten Wirkungsgeschichte.

Darin bestätigt er die Rezeption von Feigls „Existential Hypotheses" im Kraft-Kreis, nachdem dort auch der persönliche Auftritt Wittgensteins zum Problem der Außenwelt nicht überzeugen konnte (Feyerabend 1966, 4):

> It was at this stage of confusion and uncertainty that our intention was directed towards Feigl's "Existential Hypotheses". Our debates now took a completely new turn. This paper, taken together with Kraft's own contributions and with the ideas which Popper had explained to us on the occasion of his visits to the Alpbach Summer University in the summers of 1948 and 1949, greatly diminished or doubts about realism. There were still some points which were not entirely clear, and I wished for an opportunity to discuss the matter with Feigl in person. Another problem that had come up with the realism-positivism issue concerned the application of the calculus of probabilities.

Das erste Zusammentreffen von Feyerabend mit Feigl in Wien im Jahre 1954 mit der gewünschten Diskussion über das Anwendungsproblem der Wahrscheinlichkeit beeindruckte ersteren wegen dessen überzeugenden common-sense-Stils im Kontext eines philosophischen Realismus (mit Präferenzen für Popper gegenüber Wittgenstein). Es war also keine Überraschung, dass Feyerabend mindestens zweimal als Gast am Minnesota Center wirken sollte, dessen intellektuelles Klima er folgendermaßen beschrieb (Feyerabend 1966, 9):

> The atmosphere at the Center, and especially Feigl's own attitude, his humor, his eagerness to advance philosophy and to get at least a glimpse of the truth, and his quite incredible modesty, made impossible from the very beginning that subjective tension that occasionally accompanies debate and that is liable to turn individual contributions into proclamations of faith rather than into answers to questions chosen. The critical attitude was not absent, on the contrary, one now felt free to voice basic disagreements in clear, sharp, straightforward fashion. The discussions were, and still are, in many respects similar to the earlier discussion in the Vienna Circle. The differences are that things are seen now to be much more

complex than was originally thought and that there is much less confidence that a single, comprehensive empirical philosophy might emerge.

Diese allgemeine Charakterisierung verdeckt die Tatsache, wie produktiv Feyerabend in das MCPS unter der Leitung von Feigl eingebunden war, mit dem er von 1957–1968 auch korrespondierte. Dort finden sich folgende Manuskripte (vgl. „The Book", Walter Library, University of Minnesota):

1955: Carnap's Theory of Interpretation of Theoretical Systems
A Note on Carnap's New Criterion of Empirical Science
1957: Replies to Hempel and Carnap
1957: On the Quantum Theory of Measurement ... Appendix
1958: An Attempt at a Realistic Interpretation of Theoretical Knowledge Complementarity; Reply to A.Grünbaum; Some further Comments on Conventionality in Geometry (ad Reichenbach); Further Notes on Conventionalism; Comments on Rozeboom
1959: Reply to Hanson
1960: Explanation, Reduction, and Empiricism

Aus der Sicht des älteren Freundes liest sich die Bekanntschaft mit dem jüngeren Wiener Philosophen wie folgt (Feigl 1968, 668):

I met Feyerabend on my first visit to Vienna after the war (my last previous visit was in 1935). This was in the summer of 1954 when Arthur Pap was a visiting professor at the University of Vienna. Feyerabend had been working as an assistant to Pap. Immediately, during my first conversation with Feyerabend, I recognized his competence and brilliance. He is, perhaps, the most unorthodox philosopher of science I have ever known. We have often discussed our differences publicly. Although the audiences usually sided with me my more conservative views, it may well be that Feyerabend is right, and I am wrong.

Das ist vielleicht auch eine Anspielung auf Feyerabends erste Publikation von *Against Method* (15 Kapitel und 1 Anhang im Umfang von 114 Seiten im Vergleich mit der deutschen Ausgabe von 1983: 19 Kapitel mit 3 Anhängen) im Band IV der von Feigl und Maxwell herausgegebenen *Minnesota Studies for the Philosophy of Science*, deren Ursprung in der biografischen und autobiografischen Literatur von Feyerabend unerwähnt geblieben ist. Die Herausgeber des speziellen Bandes über *Analyses of Theories and Methods of Physics and Psychology*, Michael Radner und Stephen Winokur (1970), stellen den Beitrag zusammen mit Aufsätzen von Feigl, N.R. Hanson, Carl G. Hempel, Mary Hesse, Grover Maxwell, Joseph Margolis und William W. Rozeboom als durchaus „normalwissenschaftliches" Resultat von einschlägigen Konferenzen und Diskussionen am MCPS vor. Vor diesem Hintergrund verwundert es, wie verkürzt und ungenau Feyerabend über die Entstehung seines späteren Bestsellers – dessen Vorgeschichte inzwischen sehr gut in dem von Matteo Motterlini herausgegebenen Band *For and Against Method* (1999) rekonstruiert worden ist – im Vorwort der deutschen Ausgabe von *Wider den Methodenzwang* schreibt:

> Im Jahre 1970 zog mich Imre Lakatos, einer der besten Freunde, die ich je besessen hatte, zur Seite und sagte mir: „Paul", sagte er, „du hast doch so komische Ideen. Warum schreibst Du sie nicht nieder, ich schreibe eine Antwort, wir publizieren die Sache und haben einen Heidenspaß." Der Vorschlag gefiel mir und ich machte mich an die Arbeit. Das Manuskript meines Teiles meines geplanten Buches war im Jahre 1972 beendet und ich schickte es nach London. Dort verschwand es auf geheimnisvolle Weise. Imre Lakatos, der dramatische Gesten liebte, verständigte die Interpol, und in der Tat, die Interpol fand mein Manuskript und schickte es an mich zurück. Ich las es noch einmal und schrieb es zum großen Teil um. Im Februar des Jahres 1974, nur wenige Wochen nachdem ich meine Revision beendet hatte, starb Imre Lakatos. Ich habe dann meinen Teil ohne seine Antwort publiziert. (Feyerabend 1986, 11).

Die Gründe für diese Enthistorisierung mögen darin liegen, dass ab 1970 für den erfolgreichen *agent provocateur* der Wissenschaftsphilosophie der Kontext des Dritten Wiener Kreises um Kraft bis hin zum „Wiener Kreis in America" um Feigl (1968) nicht mehr opportun gewesen ist. Der Erfolg seiner – großteils berechtigten – Kritik an der analytisch-normativen Wissenschaftstheorie auch ohne die geplante Ergänzung durch den kritisch-rationalen Lakatos hat ihn offenbar angesichts seines neuen Images als einer Ikone der Postmoderne sowie des epistemologischen und kulturellen Relativismus zu diesem Weg ermutigt. Erst gegen Ende seines Lebens kehrte er wieder zurück zu den Wiener Wurzeln seiner intellektuellen Entwicklung im Sinne einer Einheit von Wissenschaftsgeschichte und Wissenschaftstheorie – zu Mach, Boltzmann, Alois Riegl u.a. bis hin zur historischen Tradition im Wiener

Kreis, auch wenn er diesen undifferenziert besonders im Zusammenhang mit Popper kritisiert.

Feyerabends philosophische Heimkehr: Ernst Mach und die historische Tradition in der Wissenschaftstheorie

Spätestens mit dem Ruf an die ETH Zürich in den 1980er Jahren kam es zur physischen und geistigen Rückkehr Feyerabends zu seiner frühen intellektuellen Sozialisation mit der erneuten Mach-Lektüre und der verstärkten Ausarbeitung seines Relativismus und Pluralismus, nachdem er schon früher (1962) das Konzept der Inkommensurabilität und seine kontextuelle Bedeutungstheorie vertreten hatte Dieser Prozess scheint parallel zu laufen mit der Herausgabe seiner beiden Bände *Philosophical Papers* (1981) bzw. *Ausgewählte Schriften* (1978 und 1981), die er ergänzte, neu bearbeitete und teilweise mit Nachworten versah. Dies setzte sich mit der Veröffentlichung seines Buches *Farewell to Reason* (1987) fort, das in deutscher Übersetzung als *Irrwege der Vernunft* (1989) erschien.

In seinem – Adolf Grünbaum gewidmeten – Aufsatz „Machs Theorie der Forschung und ihre Beziehung zu Einstein" (1988) würdigt er – übrigens ohne Referenz zu den Machianern Philipp Frank oder Richard von Mises – dessen heuristische und historische Einstellung zur Forschungsmethode (zur Interpretation vgl. Stadler 1982, 123ff. und 1988, 40f.): es geht um die Präferenz einer historisch-kritischen Forschungstheorie gegenüber der rationalistischen, abstrakt-theoretischen Tradition in der Wissenschaftsphilosophie, und er setzt die Rekonstruktion des Machschen Werkes fort, die er bereits in *Erkenntnis für freie Menschen* (1980, 273f.) programmatisch so formuliert hatte:

> Die Machsche Kritik war ein Teil der Reform der Wissenschaften, sie verband Kritik mit neuen Ergebnissen; die Positivisten aber und ihre atemlosen Gegner, die „kritischen Rationalisten", beginnen mit einigen gefrorenen Bestandteilen der Wissenschaften, die der Forschung nicht mehr zugänglich sind, und verstärken sie mit Hilfe philosophischer Argumente (Poppers „Beiträge" zum Realismus). Machs Kritik war dialektisch und fruchtbar, die Kritik der Philosophen ist dogmatisch und ohne Ergebnisse.

In dieser Tradition sieht Feyerabend übrigens auch Einstein, Bohr und Neurath, ohne die Globalkritik am Wiener Kreis zurückzunehmen. Im

Anschluss an diese Rehabilitierung versuchte Feyerabend, die Forschungstheorie Machs zwischen Abstraktion und Phantasie von seiner phänomenalistischen Erkenntnistheorie getrennt zu würdigen, wodurch die Dualismen von Theorie und Erfahrung, Philosophie und Wissenschaft, Materialismus und Idealismus sowie Wissenschaftsgeschichte und Wissenschaftstheorie überwunden werden sollten:

> Machs physikalische Argumente ergeben zusammen eine Wissenschaftsphilosophie, die sich vom Positivismus unterscheidet, mit Einsteins Forschungspraxis ... übereinstimmt und im übrigen völlig sinnvolle Einwände gegen Atomvorstellungen des 19. Jahrhunderts und gegen die Relativitätstheorie mit sich bringt ... Es wird sich auch erweisen, dass Machs „Erkenntnistheorie" gar keine Erkenntnistheorie ist. Sie ist vielmehr eine allgemeine wissenschaftliche Theorie (oder ein Theorieentwurf), die ihrer Form (nicht aber ihrem Inhalt) nach einem Atomismus vergleichbar ist, sich aber von jeder positivistischen Ontologie unterscheidet. (1988, 435f.)

Als Resumé, oder „Moral der Geschichte" dieser wissenschaftsgeschichtlichen Fallstudie empfiehlt Feyerabend (461f.) schließlich eine Skepsis gegenüber etablierten Meinungen, die Lektüre der Originalliteratur als Korrektiv, die Erkennung der Simplifizierung etablierter Meinungen sowie der „Scheingefechte" wie zwischen „Positivismus" und „Realismus", letztlich der Irreführung durch rein philosophische Systeme. Schließlich sei deren „Spiegelfechterei" ein Segen für alle jene, die Philosophen bleiben wollen, wenn sie sich nicht von den „Märchen" retten lassen wollen.

Schlussbemerkung

Lässt man die oben beschriebenen Befunde zum frühen Feyerabend der Jahre 1945–1955 Revue passieren, dann ergibt sich folgendes Bild als Ergänzung und Korrektur der bisherigen Forschung:

Die starke Fokussierung auf die Zeit nach der Buchpublikation von *Against Method* (1975) verdeckt die starke Prägung durch, und Wechselwirkung mit dem „Dritten Wiener Kreis" des Kraft-Zirkels im Wien der Nachkriegszeit sowie mit dem Österreichischen College / Forum Alpbach bis in die 1960er Jahre. Darüberhinaus ist die Kontinuität dieses intellektuellen Umfeldes im Minnesota Center for the Philosophy of Science mit Herbert Feigl als wichtigste Bezugsperson für den „Wiener

Kreis in Amerika" personell und thematisch relevant geblieben: Themen wie Materialismus/Realismus/Idealismus, Relativismus, Theorie und Erfahrung, Bestätigung und Bewährung sowie Erklärung und Prognose bildeten weiter den Gegenstandsbereich der pluralistischen Diskussionen in der Philosophy of Science.

Obwohl diese philosophische Entwicklung durch die beiden Bände von Feyerabends ausgewählten Schriften in Englisch und Deutsch gut nachvollziehbar ist, und außerdem dessen Autobiografie – wenn auch fragmentarisch und selektiv – wesentliche Elemente dieser Entwicklungsgeschichte beinhaltet, ist dieser Prozess vor allem im deutschsprachigen Bereich kaum wahrgenommen worden. Dies wurde u.a. auch durch Feyerabend selbst verstärkt, der sich angesichts des Erfolges von *Wider den Methodenzwang* (1976) und *Erkenntnis für freie Menschen* (1980) gerne als postmoderner Denker des „anything goes" und der „anarchistischen Erkenntnistheorie" feiern ließ.

Erst in der letzten Dekade seines Lebens in Zürich wurden seine philosophischen Wurzeln von Mach bis zum Wiener Kreis und der „Nordic Connection" wieder stärker und systematischer sichtbar. Die Gründe für diese subjektive Ausblendung und historiografische Lücke mögen einerseits in der heftigen bis polemischen Abkehr von Poppers Kritischem Rationalismus, andererseits in der verkürzten Wahrnehmung des Logischen Empirismus (als einer ahistorischen analytischen Wissenschaftstheorie) gelegen sein. Allein die Kenntnis des nichtreduktiven Naturalismus und Relativismus von Otto Neurath und Philipp Frank, dessen Position in Zeiten des Kalten Krieges deswegen heftig kritisiert worden war, hätte diese Opposition gerade angesichts der Wiederentdeckung von Ernst Mach mildern müssen, auch wenn die späte Frontstellung zum Rationalismus und zur post-aufklärerischen Wissenschaftsphilosophie als Differenz bestehen bleibt. All dies beeinträchtigt keineswegs die unbestrittene Originalität und intellektuelle Eigenständigkeit des so oft missverstandenen kreativen Forschers und wachen demokratischen Bürgers Paul Feyerabend.

Insofern scheint es nicht übertrieben, Paul Feyerabend als einen Philosophen aus Wien und der Wiener Kreise zu bezeichnen, der durch diese Philosophie (vom ersten bis zum dritten Wiener Kreis inkl. Wittgenstein und Popper, vom „Anderen Zauberberg" bis zum „Minnesota Center") maßgeblich beeinflusst worden ist – und der gegen Ende seines Lebens wieder konstruktiv und provokant zugleich zu diesen philosophischen Traditionen zurückgekehrt ist.

In seinem letzten Interview gibt Feyerabend (1994) selbst auf die Frage, warum er nicht nach Wien zurückgekehrt sei und ob er dort schlechte Erfahrungen gemacht habe, folgende für ihn typische Antwort:

> Nein, nein. Ich habe dort ausgezeichnete Erfahrungen gemacht, es war wundervoll. Es gab dort sehr gute Leute. Wissen Sie, wir Physikstudenten fielen in die philosophischen Vorlesungen ein, standen mittendrin auf und sagten: „Das ist doch alles Unsinn, worüber hier geredet wird". Dann wurden wir aus den Vorlesungen geworfen.

Was der Interviewte hier vornehm verschweigt, ist die skandalöse Tatsache, dass er sowohl als hochbegabter Jungwissenschaftler als auch als international renommierter Philosoph nie zur Rückkehr an die Universität nach Wien eingeladen wurde. So blieb es eben bei einer geistigen Rückkehr, deren Potenzial und Erträge noch immer zu würdigen sind.

Persönliche Nachbetrachtung

Der Autor dieses Beitrages hat Feyerabend leider nicht persönlich kennengelernt, hatte aber das große Vergnügen, ihn als Vortragenden im Wiener Alten Rathaus noch erlebt zu haben. Zuvor hat er mir eine nette Postkarte gesandt, nachdem ich ihm mein Buch über die Wirkungsgeschichte von Ernst Mach (1982) – mit Erwähnung seiner eigenen Mach-Rezeption – geschickt hatte. Darin schrieb er in seinem bekannt launigen Stil und mit einer Vorwegnahme der hier vorgebrachten Thesen:

> Lieber Herr Stadler – vielen Dank für Ihre schöne Tratschkiste – ich meine Ihren Bericht über die Positivisten, ihre Vorläufer und Nachfahren. Ich habe nur da und dort hineingeschaut und mich gefreut, so viele bekannte Namen gefunden zu haben, vor allem meine Lehrer Thirring, Kraft und Hollitscher (DIE waren meine Lehrer und nicht dieser Windbeutel, der Popper). Ein weiterer Lehrer, leider auch zu meiner Studienzeit schon tot war aber der Mach und zu diesem lege ich als überzeugter Mach Epigone eine weitere Exegese an, die so ziemlich allem zuwiderläuft, was gelehrte Herren wie der Holton und andere gegenwärtig über den Mach sagen (der Gereon Wolters aber stimmt mir zu). Ja, und vielleicht sehen wir uns doch einmal irgendwo (gez. Paul Feyerabend).

Anmerkung: Die angesprochene „weitere Exegese" waren die beigelegten Fahnen von Feyerabends oben zitiertem Aufsatz „Mach's Theory of Research and its Relation to Einstein", (zuerst in: *Studies in the History and Philosophy of Science*, Vol. 15, No.1). Der Beitrag erschien auch in *Farewell to Reason* (1987).

Autorisierte deutsche Übersetzungen finden sich in: Rudolf Haller / Friedrich Stadler (Hrsg.), *Ernst Mach – Werk und Wirkung*. Wien: Hölder-Pichler-Tempsky, 1988, S. 435-462 und in Feyerabends *Irrwege der Vernunft*. Frankfurt/M.: Suhrkamp 1989. Mit einem Zusatz 1988, S. 277-311.

* * *

Der Autor hat die auf Feyerabend bezogene Forschung erstmals mit seiner Dissertation über Ernst Mach (publiziert 1982) begonnen. Dem folgte ein Kapitel im Rahmen des Endberichts zum Forschungsprojekt *Wissenschaftslogik – Philosophy of Science – Wissenschaftstheorie* (Wien: Bundesministerium für Wissenschaft und Verkehr 1997).

Darauf aufbauend und in Weiterführung meiner bis 1934 bzw. 1938 reichenden Monografien *Vom Positivismus zur „Wissenschaftlichen Weltauffassung"* (Wien–München: Löcker 1982) und *Studien zum Wiener Kreis* (Frankfurt/M.: Suhrkamp 1997 / Englisch: Springer: Wien–New York 2001) wurden inzwischen als weitere diesbezügliche Bausteine folgende Beiträge publiziert:

Friedrich Stadler, „Philosophie – Zwischen ‚Anschluss' und Ausschluss, Restauration und Innovation", in: Margarete Grandner, Gernot Heiss, Oliver Rathkolb (Hrsg.), *Zukunft mit Altlasten. Die Universität Wien 1945 bis 1955*. Innsbruck–Wien–München–Bozen: StudienVerlag, 2005, 121-136.
Friedrich Stadler, „La contribution du retour d'exil des intellectuels à la culture scientifique de la Seconde République", in: *Exil et retour d'exil*. Etudes réunies par Paul Pasteur et Friedrich Stadler. Rouen 2004, 37-54. (= *Austriaca 56/2003*).
Friedrich Stadler, "The 'Wiener Kreis' in Great Britain: Emigration and Interaction in the Philosophy of Science", in: Edward Timms / Jon Hughes (Eds.), *Intellectual Migration and Cultural Transformation. Refugees from National Socialism in the English Speaking World*. Wien–New York: Springer, 2003, 155-180.
Friedrich Stadler, "Transfer and Transformation of Logical Empiricism: Quantitative and Qualitative Aspects", in: Gary Hardcastle / Alan

Richardson (Eds.), *Logical Empiricism in North America*. Minneapolis: University of Minneapolis Press 2003, 216-233.

Friedrich Stadler, „Zur politischen Relevanz und zum kulturellen Kontext des Logischen Empirismus", in: Michael Heidelberger / Friedrich Stadler (Hrsg.), *Wissenschaftsphilosophie und Politik / Philosophy of Science and Politics*. Wien–New York: Springer, 2003, 9-24.

Friedrich Stadler, „Wendung und Verblendung – Walter Hollitscher zwischen Wiener Kreis und Marx", in: Alfred Klahr Gesellschaft (Hrsg.), *Zwischen Wiener Kreis und Marx. Walter Hollitscher (1911–1986)*. Wien 2001, 59-63.

Friedrich Stadler 2006: Organizer of the Symposium „On the Banishment and Return of the Philosophy of Science after World War II" im Rahmen von HOPOS 2006, 6th International History of Philosophy of Science Congress, in Paris, 14.-18. Juni 2006. Vgl. Abstracts, pp.82-86.

Im Druck:

„History of Philosophy of Science" in: Theo Kuipers (ed.), *Handbook of Philosophy of Science – Focal Issues*. Elsevier.

"The Vienna Circle. Context, Profile, and Development", in: Alan Richardson / Thomas Uebel (Eds.), *The Cambridge Companion of Logical Empiricism*. Cambridge University Press.

Anmerkung

[*] Die vorliegende Studie ist Teil eines umfangreicheren Forschungsprojektes zur Geschichte der Wissenschaftstheorie im 20. Jahrhundert im Zusammenhang mit einem internationalen Projekt am Beispiel von Rudolf Carnap und Wolfgang Stegmüller am Institut Wiener Kreis, das vom österreichischen Forschungsfonds (FWF) finanziert wird. (vgl. die Projektbeschreibung unter: http://www.univie.ac.at/ivc/stegmueller/). Besonderer Dank gilt hier dem Projektmitarbeiter Mag. Christoph Limbeck-Lilienau für seine einschlägigen Archivarbeiten und Literaturrecherchen im Archiv des Österreichischen College in Wien und dem Archiv mit dem Feyerabend-Nachlass an der Universität Konstanz. Beiden Einrichtungen sei ebenfalls für die gute Zusammenarbeit und Benützung der Bestände gedankt.

Literaturverzeichnis

Hans Albert, „Wissenschaft in Alpbach", in: Alexander Auer (Hrsg.), a.a.O. *Forum Alpbach*, 17-22.
Alexander Auer (Hrsg.), *Das Forum Alpbach 1945–1994*. Wien: Ibera Verlag-European University Press 1994.
Aufklärung und Kritik. Sonderheft 8/2004. Schwerpunkt: Ernst Topitsch.
Hans Peter Duerr (Hrsg.), Versuchungen. Aufsätze zur Philosophie Feyerabends. 2 Bände. Frankfurt/M.: Suhrkamp 1980/81.
Herbert Feigl, *Inquiries and Provocations. Selected Writings*. Ed. by Robert S. Cohen. Dordrecht–Boston–London: Reidel 1981.
Herbert Feigl, "Existential Hypotheses. Realistic Versus Phenomenalistic Interpretations", in: *Philosophy of Science* 17/1950, 35-62. Auch in: *Ebda.*, 192-223.
Herbert Feigl, „The *Wiener Kreis* in America" 1968/69, in: *Ebda.*, 57-94.
Paul Feyerabend, *Zur Theorie der Basissätze*. Unveröffentlichte Dissertation. Universität Wien 1951.
Paul Feyerabend, Die Geisteswissenschaften in Österreich. Unveröffentlichtes Manuskript. Wien 1954. (163pp).
Paul Feyerabend, "Wittgenstein's Philosophical Investigations" in: *The Philosophical Review* 64/1955, 449-483. Auch in Feyerabend 1981, 293-325.
Paul Feyerabend, „Das Problem der Existenz theoretischer Entitäten", in: Topitsch (Hrsg.), a.a.O., 35-72.
Paul Feyerabend, "Against Method: Outline of an Anarchistic Theory of Knowledge", in: Michael Radner / Stephen Winokur (Eds.), *Analyses of Theories and Methods of Physics and Psychology*. Minneapolis: University of Minnesota Press 1970, 17-130. (= Minnesota Studies in the Philosophy of Science, Vol. IV, ed. by Grover Maxwell and Herbert Feigl).
Paul Feyerabend / Grover Maxwell (Eds.), *Mind, Matter, and Method. Essays in Philosophy and Science in Honor of Herbert Feigl*. Minneapolis: University of Minnesota Press 1966.
Paul Feyerabend, "Herbert Feigl: A Biographical Sketch", in: *Ebda.*, 3-16.
Paul Feyerabend, *Against Method. Outline of an Anarchistic Theory of Knowledge*. London: New Left Books 1975.
Paul Feyerabend, *Science in a Free Society*. London: New Left Books 1978.

Paul Feyerabend, *Erkenntnis für freie Menschen*. Frankfurt/M.: Suhrkamp 1980.
Paul Feyerabend, *Wider den Methodenzwang. Skizze einer anarchistischen Erkenntnistheorie*. Frankfurt/M.: Suhrkamp 1976. Revidierte und erweiterte Fassung 1983.
Paul Feyerabend, *Der wissenschaftstheoretische Realismus und die Autorität der Wissenschaften. Ausgewählte Schriften*, Band 1. Braunschweig–Wiesbaden: Vieweg & Sohn 1978.
Paul Feyerabend, *Probleme des Empirismus. Schriften zur Theorie der Erklärung, der Quantentheorie und der Wissenschaftsgeschichte. Ausgewählte Schriften*, Band 2. Braunschweig–Wiesbaden: Vieweg & Sohn 1981.
Paul Feyerabend, *Irrwege der Vernunft*. Aus dem Amerikanischen von Jürgen Balsius. Frankfurt/M.: 1989.
Paul Feyerabend, *Zeitverschwendung*. Übersetzt von Joachim Jung. Frankfurt/M.: Suhrkamp 1995.
Paul Feyerabend, Die Vernichtung der Vielfalt. Ein Bericht. Aus dem Englischen von Volker Bönig und Rainer Noske. Hrsg. Von Peter Engelmann. Wien: Passagen Verlag 2005.
Kurt R. Fischer / Franz Wimmer (Hrsg.), *Der geistige Anschluß. Philosophie und Politik an der Universität Wien 1930–1950*. Wien: WUV Verlag 1993.
Philipp Frank, *Modern Science and its Philosophy*. Cambridge: Harvard University Press 1949.
Philipp Frank, *Wahrheit – relativ oder absolut?* Mit einem Vorwort von Albert Einstein. Zürich: Pan-Verlag 1952.
Margaret Grebowicz 2006: Organizer of the Symposium *"Against Method*: Thirty Years Later." Abstracts of the 6th International History of Philosophy of Science Congress, Paris, 14.-18. Juni 2006. *Abstracts*, 64-67.
Hans Grümm, *Drei Leben. Krieg, Partei, Atom*. Wien: Löcker 1992.
Rudolf Haller, „In memoriam Paul Feyerabend", in: Friedrich Stadler (Hrsg.), *Bausteine Wissenschaftlicher Weltauffassung*. Wien–New York: Springer 1997, 91-100.
Ernst Hanisch, *Der lange Schatten des Staates. Österreichische Gesellschaftsgeschichte im 20. Jahrhundert*. Wien: Ueberreuter 1995.
Willy Hochkeppel, „Propheten im eigenen Lande ... Über den Umgang mit einem großen philosophisch-wissenschaftlichen Erbe in der 2. österreichischen Republik. Impressionen eines Außenseiters", in: Gerbert Frodl / Paul Kruntorad / Manfried Rauchensteiner (Hrsg.), *Physiognomie der 2. Republik von Raab bis Bruno Kreisky*. Wien 2006, 115-151.

Gerald Holton, *Thematische Analyse der Wissenschaft. Die Physik Einsteins und seiner Zeit.* Frankfurt/M.: Suhrkamp 1981.
Paul Hoyningen-Huene, "Paul Feyerabend", in: *Journal for General Philosophy of Science* 1997/28, 1-18.
Paul Hoyningen-Huene, „Paul Feyerabend – ein postmoderner Philosoph?" in: *Information Philosophie* 1/2002, 30-37.
Paul Hoyningen-Huene, „Paul Feyerabend und Thomas Kuhn", in: *Journal for the General Philosopy of Science* 33/2002, 61-83.
Alfons Keupink / Sanford Shie (Eds.), *The Limits of Logical Empiricism. Selected Papers of Arthur Pap.* Dordrecht: Springer 2006.
Alfred Klahr Gesellschaft (Hrsg.), *Zwischen Wiener Kreis und Marx. Walter Hollitscher (1911–1986).* Wien 2001.
Viktor Kraft, *Die Grundformen der wissenschaftlichen Methoden.* (= Sitzungsberichte der Österreichischen Akademie der Wissenschaften, philos.-histor. Klasse 203). Wien 1925, 1-104.
Viktor Kraft, *Der Wiener Kreis. Der Ursprung des Neopositivismus.* Wien–New York: Springer 1950. 3. Auflage 1997.
Otto Molden, *Der andere Zauberberg. Das Phänomen Alpbach.* Wien–München–Zürich–New York 1981.
Matteo Motterlini (Ed.), *For and Against Method. Imre Lakatos and Paul Feyerabend.* Including Lakatos's Lectures on Scientific Method and the Lakatos-Feyerabend Correspondence. Chicago–London: The University of Chicago Press 1999.
Matteo Motterlini, "Paul Karl Feyerabend", in: Sahotra Sarkar / Jessica Pfeifer (Eds.), *The Philosophy of Science. An Encyclopedia.* New York–London: Routledge, 304-310.
Arne Naess, *Erkenntnis und wissenschaftliches Verhalten.* Oslo 1936.
Arne Naess, „Paul Feyerabend – Ein Held der Grünen?", in: Hans-Peter Dürr (Hrsg.), a.a.O., 184-199.
Österreich – Geistige Provinz? Wien–Hannover–Bern: Forum Verlag 1965.
Arthur Pap, *Analytische Erkenntnistheorie. Kritische Übersicht über die Entwicklung in USA und England.* Wien: Springer Verlag 1955.
John Preston, *Feyerabend. Philosophy, Science and Society.* Cambridge: Polity Press 1997.
Willard Van Orman Quine, "Two Dogmas of Empiricism", in: *Philosophical Review* 60, 1951, 20-43.
Jan Radler, *Victor Krafts konstruktiver Empirismus. Eine historische und philosophische Untersuchung.* Berlin: Logos Verlag 2006.
George Reisch, *How the Cold War Transformed Philosophy of Science.* Cambridge University Press 2004.

Wolfgang Stegmüller, *Rationale Rekonstruktion von Wissenschaft und ihrem Wandel.* Mit einer autobiografischen Einleitung. Stuttgart: Reclam 1979.

Ernst Topitsch (Hrsg.), *Probleme der Wissenschaftstheorie. Festschrift für Victor Kraft.* Wien: Springer Verlag 1960.

Gereon Wolters, *Mach I, Mach II, Einstein und die Relativitätstheorie.* Berlin–New York: de Gruyter 1987.

KURT RUDOLF FISCHER

FEYERABENDS WELTANSCHAUUNG*

Mein Beitrag wurde angekündigt sowohl als von einem, der eine persönliche Beziehung zu Paul Feyerabend hatte als auch als ein Beitrag, der sich mit seiner Weltanschauung beschäftigt. Über ersteres Thema habe ich bereits berichtet, werde aber heute noch einiges hinzufügen und wiederholen und mich auch mit Feyerabends Weltanschauung beschäftigen.[1] Zuerst möchte ich aber etwas zitieren: „Weltanschauung ist, besorge ich, ein spezifisch deutscher Begriff", meint Sigmund Freud in seiner *Neuen Folge der Vorlesungen in die Psychoanalyse*.[2] In einem vom Institut Wiener Kreis veranstalteten und geförderten Symposion den Begriff „Weltanschauung" zu verwenden ist fast eine Provokation. Im Wiener Kreis war doch die Weltanschauung eher etwas Emotives, ein Ausdruck von Gefühlen, aber doch keine intellektuelle oder wissenschaftliche Angelegenheit und schon gar keine wissenschaftstheoretische. Freuds weitere Aussage darüber, was man unter einer Weltanschauung versteht, ist, dass „eine Weltanschauung eine intellektuelle Konstruktion ist, die alle Probleme unseres Daseins aus einer übergeordneten Annahme löst."[3] Die Psychoanalyse hält er für einen Zweig der Wissenschaft Psychologie. Freud zu zitieren bei einem Vortrag über Feyerabend ist auf jeden Fall problematisch, weil dieser kein besonderer Freund der Psychoanalyse war. Er erzählte mir einmal, dass er in London einen psychoanalytischen Psychiater aufgesucht hätte. Er verwickelte diesen aber in eine solche Diskussion, dass der Psychiater seine Praxis vorübergehend aufgeben musste und Urlaub nahm. Natürlich war der Psychiater nicht in der Lage gewesen, ihm zu helfen.

Der Logische Positivismus oder Logische Empirismus des Wiener Kreises, von dem, grob gesprochen, Feyerabend seinen wissenschaftstheoretischen Anfang nahm, wäre wohl mit dem Zitat von Freud einverstanden gewesen. Es gab eine Verwandtschaft zwischen dem Wiener Kreis und der Psychoanalyse. Beide waren durch staatliche Macht von Antisemitismus und Faschismus verfolgt, zuerst im Ständestaat und dann im Dritten Reich.[4] Aber es gab auch personelle Verbindungen, z.B. durch Egon Brunswik und Else Frenkel-Brunswik.[5]

In früheren Zeiten war eine Weltanschauung wichtiger Ausdruck einer geistigen Grundhaltung. Die ideologische Auseinandersetzung zwischen Religion und Wissenschaft spielte dabei eine bedeutende

Rolle. Als ich in Shanghai mein Studium an der St. John's University begann, einem College, das von der amerikanischen Episcopelian Mission gegründet und finanziert wurde, gab es eine Reihe von Lehrveranstaltungen, die zeigen sollten, dass Religion und Wissenschaft kompatibel sind. Das war im Jahre 1944. Vorher noch, als Halbwüchsiger in Wien, dachte ich, dass das Studium der Medizin und der Philosophie mir eine Weltanschauung geben würde, durch die ich eine solide geistige Basis für mein ganzes Leben bekommen könnte. Ich war bei meinen weiteren Studien in Berkeley nicht wenig erstaunt, dass man sich durch die Philosophie nicht Grundsätze und Resultate aller Disziplinen erwerben konnte sondern dass man im Gegenteil zumindest eine Disziplin studiert haben musste und über diese und mittels dieser philosophieren konnte.[6]

Paul Feyerabend gehörte derselben Generation an wie ich – er war zwei Jahre jünger –, und seine weltanschauliche Basis war ihm wohl von den Eltern vermittelt worden. Sie war streng katholisch, überhaupt streng. Einmal erzählte er mir, dass er eine Ohrfeige bekam, wenn er seine Hände in den Hosentaschen hatte, denn es wurde vermutet, er würde onanieren. In seiner Autobiographie schreibt er:

> Ich glaubte fest an Engel und Geister. Sie konnten überall auftauchen und waren zu allem imstande. Ich fürchtete Gott. Er war weit weg und ziemlich farblos, er existierte nur, aber er war auch mächtig und kannte die geheimsten Dinge.[7]

Aber die große und bleibende Bedeutung von Weltanschauung wurde mir erst dann besonders klar, als ich Ende der siebziger Jahre des vorigen Jahrhunderts nach Wien zurückkam und Freunde zum Heurigen einlud, die mit ihren Kindern kamen. Die einen hatten zwei Mädchen im Alter von zehn und zwölf Jahren, die anderen zwei Buben im gleichen Alter. Eine Woche später fragte ich die Mädchen, wie ihnen die Buben gefallen hätten. „Gar nicht", war die Antwort, denn sie hätten ja nicht einmal eine Weltanschauung gehabt.

Feyerabend ist während seiner Schulzeit auf die Philosophie gestoßen. Er schreibt in seiner Autobiographie, dass er antiquarische Taschenbücher gekauft hatte – die Eltern erlaubten ihm nicht, aus dem Haus zu gehen –, wobei „hin und wieder auch ein Band von Platon, Descartes oder Büchner (dem Materialisten, nicht dem Dichter) darunter waren."[8] In der Schule fiel er auf. Sein Deutschlehrer, Professor Wiener (er war übrigens auch mein Deutschlehrer) nahm sich seiner an und korrigierte seine Gedichte. Dem Physikprofessor Thomas verdankte Feyerabend sein Interesse für Astronomie und Physik, Gegenstände, die er zusammen mit Mathematik studieren wollte.

Es sei hier kurz die Philosophie skizziert, mit der sich Feyerabend identifizierte, nachdem er seine ursprüngliche Absicht aufgegeben hatte, sich nach dem Krieg den Geisteswissenschaften zuzuwenden, besonders der Geschichte, denn er wollte verstehen, was eigentlich vorgefallen war. Die meisten Leser werden mit der Philosophie vertraut sein, auf die Feyerabend nach dem Krieg stieß. Im Theologie-Seminar von Pater Otto Mauer, das Feyerabend besuchte, rief er mehrmals aus, dass die Wissenschaft die Grundlage allen Wissens sei, und dass nicht-empirische Überlegungen entweder Logik oder Unsinn seien.[9] Diese Ansicht war grundlegend für den Wiener Kreis. Der Wiener Kreis hatte sich in den dreißiger Jahren aufgelöst, aber in den vierziger Jahren die philosophische Herrschaft in den angelsächsischen Ländern angetreten.

Historisch betrachtet gehörte der Wiener Kreis in eine Tradition, die Rudolf Haller als österreichische Philosohie identifizierte, mit den drei Merkmalen: Empirismus, Wissenschaft als bevorzugte Weise der Erkenntnis und Sprachanalyse als einzige Aufgabe der Philosophie.[10] Bekannt und besonders einflussreich wurde diese Philosophie durch ein Büchlein von Alfred Ayer, einem Engländer, der sich studienhalber einige Zeit in Wien aufgehalten hatte. Der Titel seines Buches war *Language, Truth and Logic*.[11] In diesem wurde zwischen kognitiven und nicht-kognitiven Sätzen unterschieden. Erstere wurden in synthetische Urteile aposteriori einerseits und in analytische Urteile apriori andererseits eingeteilt. Letztere, nicht kognitive, waren auch Werturteile und drückten ein Gefühl aus oder gaben eine Stellungnahme bekannt. Die intellektuelle Stoßkraft des kleinen Buches war stark. Etwa von da ab, kann man sagen, waren Logik, Sprache und die Wissenschaften Hauptthemen philosophischer Beschäftigung zumindest an den amerikanischen und britischen Eliteuniversitäten.

Als ich als „Undergraduate" im Jahr 1949 das Studium der Philosophie in Berkeley fortsetzte, mit einem Baccalaureat abschloss und ab 1951 als Graduate weiterführte, wurde lediglich der Logische Positivismus ernst genommen, doch meist bekämpft. Moritz Schlick, das Haupt des Wiener Kreises, war als Gastprofessor in Berkeley gewesen und hatte einen starken Eindruck hinterlassen. Im Seminarraum, der auch die Institutsbibliothek war, hatte er auf die Regale gezeigt und gemeint, dass die meisten dieser philosophischen Werke nur Unsinn enthielten und kein Wissen.

Im Laufe der Zeit traten scheinbar unüberwindliche Schwierigkeiten in der Formulierung des Herzstücks des Logischen Positivismus, dem Verifikationsprinzip, auf. Das Verifikationsprinzip besagt, dass nur dann eine Aussage über die Welt, die Wirklichkeit vorliegt, wenn die Bedingungen angegeben werden können, unter denen sie als wahr gilt, sie also verifizierbar ist. Etwa zur gleichen Zeit lag Quines Aufsatz vor, in dem die grundlegende Unterscheidung des Logischen Positivismus angegriffen wurde, die Unterscheidung zwischen analytischen und synthetischen Sätzen.[12] In der Wissenschaftsphilosophie setzte sich ein Jahrzehnt später sogar eine neue Anschauung durch, die revolutionär war, Thomas Kuhns *The Structure of Scientific Revolutions*.[13] Ich hatte das Glück, zu einer kleinen Feier anlässlich dieser Publikation, die in einer Privatwohnung stattfand, eingeladen zu werden. Ich fragte Kuhn, ob er ein Logischer Positivist sei, da seine Monographie doch im Rahmen der *Encyclopedia of Unified Science* erschienen war. Kuhn lächelte, gab aber keine Antwort.

Dass eine große Kluft zwischen den beiden deutschsprachigen Philosophien besteht, der deutschen und der österreichischen, wird man wohl zugeben müssen. Erstere hatte ihre Höhepunkte in Hegel und Schelling und in Adorno und Heidegger. Sie präsentiert sich als eine Art von Übererkenntnis und stellt eine Grundlage für Leben und Wissen bereit. Der Professor sprach sowohl metaphorisch als auch tatsächlich auf die Zuhörerschaft von oben hinunter. Diskussion gab es keine. Ein Blick in das Vorlesungsverzeichnis, auch noch im zwanzigsten Jahrhundert, z.B. aus der Zeit, in der ich wieder für ein Jahr als Student in Wien war, zeigt, dass die Wiener Lehrkanzeln von Professoren der deutschen philosophischen Tradition besetzt waren. Die Philosophie stand an erster Stelle im Vorlesungsverzeichnis, vor allen anderen Fächern. In Analogie zu Rudolf Hallers Charakterisierung der österreichischen Philosophie kann man die deutsche Philosophie als idealistisch, religiös, aber auch als sprachlich bezeichnen – sprachlich allerdings in dem Sinn, dass die Sprache das Denken nicht verführt sondern Einsichten vermittelt.

Aber zurück zu Feyerabend. Im Studienjahr 1954/55 habe ich ihn das erste Mal gesehen oder, wie man bei ihm sagen kann, erlebt. Das war im Privatissimum von Professor Erich Heintel. Er wandte sich mit der Bitte an den Professor, das Privatissimum besuchen zu dürfen, was für einen, der schon das Doktorat hatte, nicht üblich war. Später meinte Feyerabend, dass er die Lehrveranstaltung nur besuchen wollte, weil der Raum geheizt war und er sich die Heizung in seiner Wohnung nicht leisten konnte. Heintel erlaubte ihm zu bleiben, wenn er den Mund hielte. Ich war empört, kam aber später zu der Erkenntnis, dass Heintel wohl wusste, warum er ihm den Mund verbot, denn Feyerabend konnte eine ganze Veranstaltung stören. Dann habe ich Feyerabend noch einmal bei einer Aufführung von *Tristan und Isolde* und bei einem Vortrag von ihm im Institut für Wissenschaft und Kunst getroffen. Man hatte damals Leute eingeladen, die von der sehr konservativen Fakultät der Wiener Universität abgelehnt wurden.

Im Sommer 1955 kehrte ich wieder nach Berkeley zurück, und im Studienjahr 1958/59 kam Feyerabend als Visiting Associate Professor nach Berkeley. Ich stellte mich ihm vor, und wir fanden heraus, dass wir beide das Robert-Hamerling-Gymnasium im achten Wiener Gemeindebezirk besucht hatten. Manche Professoren und ihre Eigenarten kannten wir beide, wir hatten also einige Gemeinsamkeiten.

Zu Feyerabends Bestellung nach Berkeley kam es folgendermaßen: Berkeley suchte einen Wissenschaftstheoretiker. Man wandte sich an Herbert Feigl, der ein jüngeres Mitglied des Wiener Kreises gewesen war und nunmehr als Direktor des Minnesota Center for the Philosophy of Science wirkte. Man hatte Feigl schon frühzeitig geraten, in die USA zu gehen, denn er war Jude und tschechoslowakischer Staatsbürger, wodurch seine Aussichten auf eine Position in Österreich schlecht waren. Feigl wurde von Alfred Tarski, dem Gründer der wissenschaftlichen Semantik und Professor der Mathematik in Berkeley, empfohlen. Feigl wiederum empfahl Feyerabend, den er wohl in Alpbach kennen gelernt hatte. Tarski hatte großen Einfluss im Department of Philosophy und unterstützte Feigl mit der Aussage, dass es nicht der Fall sei, dass Feigl immer Unsinn rede. Diese vorsichtige und mit Präzision vorgetragene Empfehlung genügte dem Department of Philosophy, Feyerabend, der im Jahr davor an der University of Bristol gelehrt hatte, einzuladen. Man war nun sicher, dass Feyerabend aus dem richtigen philosophischen Stall kam und der einzigen philosophischen Richtung angehörte, die ernst genommen werden konnte. Feyerabend war zur Zeit seiner Berufung wohl kein klassischer Logischer Positivist aber doch Mitglied des von Rudolf Haller später so genannten Wiener Mini-Kreises, dessen Haupt Viktor Kraft war.[14] Feyerabend passte gut in das Department of Philosophy, wenn wir von seiner Aversion gegen die akademische Bürokratie und deren Wichtigkeit absehen.

Das Department of Philosophy bestand zum einen Teil aus Mitgliedern, die noch vor dem Siegeszug der Analytischen Philosophie angestellt worden waren, zum anderen Teil aus Professoren, die mit der mathematischen Logik genügend vertraut waren, um sich mit der „neuen Philosophie", wie Reichenbach sie genannt hatte bzw. mit der

Philosophie aus dem Cambridge eines Moore, Russell oder Wittgenstein auseinandersetzen zu können.[15] Oft führte diese Auseinandersetzung zu einer Verwerfung der Philosophie als ernst zu nehmendes Fach überhaupt und zu einer Art von Negativismus.[16]

Feyerabend verachtete die Logik. Er wollte Erkenntnis haben, nicht Klärung. Es interessierte ihn kaum, ob die Darstellung der Erkenntnis sauber und sorgfältig war. Er sagte dies oft und wies darauf hin, wie wenig wir von der Welt eigentlich wissen. Er betrachtete die Logik als eine Art Zeitvertreib, als intellektuelle Trödelei ganz ohne Resultate. Er riet mir, mich mit jenen Teilen der Mathematik zu beschäftigen die für die Physik nützlich seien.

Für einen Philosophen, ja sogar für einen Wissenschaftstheoretiker, verstand Paul Feyerabend sehr viel von der Physik. Bei der Bestätigung seiner Kenntnisse, die vom Department of Philosophy verlangt wurde, kam es jedoch zu einem Missverständnis. Das Department of Physics bestätigte, dass Feyerabend etwa zwei Jahrzehnte hinter dem Stand der Forschung zurück lag. Professor Karplus hatte aber schriftlich *nicht* hinzugefügt, dass Wissenschaftstheoretiker meist ein halbes Jahrhundert hinter dem Stand der Forschung zurückliegen. Mit dieser Beurteilung des Physikers war Feyerabend zufrieden (und dann auch das Department of Philosophy) und er meinte, jedenfalls im Gespräch mit mir, dass kein wesentlicher Unterschied zwischen einem guten Philosophen und einem Physiker bestünde und dass er, Feyerabend, nur zu dumm sei, um ein Physiker zu sein. Außer seiner Blödheit trenne ihn eigentlich nichts davon, ein Physiker zu sein. Er erinnerte sich fälschlicherweise, dass der Physiker Thirring Zweitbegutachter seiner Dissertation gewesen wäre. Der Zweitbegutachter war aber Friedrich Kainz, Professor der Philosophie, wovon man sich leicht überzeugen kann, wenn man die Dissertation aushebt.

Feyerabend schwankte zwischen Wissenschaftstheorie, einer Disziplin, die Ordnung in die Wissenschaften bringen will, kann man wohl sagen, und Wissenschaftsgeschichte. Physik war nach dem Zweiten Weltkrieg *die* Paradewissenschaft. Ich erinnere mich noch, wie die Physikstudenten, die im International House in Berkeley wohnten, sich einigermaßen arrogant verhielten. Sie hatten für ihre Mahlzeiten einen eigenen Tisch und ließen niemanden, der nicht Physik als Hauptfach hatte, sich zu ihnen setzen. Aber auch in anderen Fächern, z.B. in der Psychologie, spielten der Logische Positivismus und die Physik eine Rolle. Der bereits erwähnte Professor Egon Brunswik schrieb, dass sein Konzept einer Wissenschaft von der Psychologie eigentlich auch der Physik entspreche, wenn man daran denkt, dass *Organismen* die

Objekte der Psychologie sind, und legte dies in seinem Beitrag zur *Encyclopedia of Unified Science* dar.

In einem Brief an Hans Albert schreibt Feyerabend, er sei nicht daran interessiert, jemanden zu beeinflussen.[17] Ich aber empfand ihn als besonders unangenehm, ja unerträglich, wenn er mich, was oft der Fall war, von etwas überzeugen wollte. Ich konnte sein schnelles Sprechen und seine drängende Art, sich in Szene zu setzen nicht leiden, auch konnte ich natürlich nicht so rasch denken wie er und intellektuell mit ihm Schritt halten. Damals bewunderte ich den feinen angelsächsischen Diskussionsstil und jene, die ihn praktizierten, besonders die Ordinary-Language-Philosophen aus England. Paul verachtete sie, auch ihre arrivierten Vertreter, als dekadent und ignorant. Doch als der bedeutendste von ihnen, John L. Austin, aus Oxford als Gastprofessor nach Berkeley kam und Lehrveranstaltungen abhielt, wurde Paul, der sich ihm in der Diskussion stellte, auf seine provokante Frage hin vernichtend abgefertigt. Austin war nicht nur noch schneller und einfallsreicher als Feyerabend, in Oxford hatte er natürlich auch mehr Übung in mündlicher Argumentation erworben. Paul änderte seine Ansichten, jedenfalls über Austin, und meinte, dass Austin eigentlich ein Wissenschafter sei. Ich glaube, Austin wäre mit dieser Beurteilung einverstanden gewesen. Auch die Sprache, das Thema der Ordinary-Language-Philosophy, hielt Feyerabend ebenso wie die Logik für nicht so wichtig. Er fürchtete, man würde über der Sprache die Natur und die Welt, von der wir noch so wenig wissen, vergessen und glauben, wichtige Probleme gelöst zu haben.

Unsere Freundschaft kam nach etwa zwei Jahren zu einem jähen Ende. Wir hatten einander fast jeden Tag getroffen oder zumindest miteinander telefoniert. Eines nachmittags, auf dem Weg zu einer Party, besuchte ich Paul. Es ging ihm gesundheitlich nicht gut: er lag mit Schmerzen und schlecht gelaunt im Bett. Ich konnte ihn dazu überreden, mit mir zu der Party zu gehen. Dort angekommen, zog er es vor, sich selbst vorzustellen: „My name is Paul Feyerabend and I am between affairs", sagte er laut. Gar bald machte er die Bekanntschaft einer jüngeren Frau – übrigens eine meiner Studentinnen –, die er dann täglich traf. Nach jeder Verabredung, um 23 Uhr herum, rief er mich immer an und wollte meinen Rat. So ging das einige Wochen lang. Meine Frau warnte mich und bat mich, ihm keine Ratschläge zu erteilen. Ich hatte keine Fertigkeiten auf diesem Gebiet und gab ihm in naiver Weise eine Menge wohlmeinender Ratschläge. Ich sagte eben alles was mir zu seinem Verhältnis einfiel und bewertete dabei auch seine Geliebte. Eines morgens um sechs Uhr erhielt ich ein Telegramm von Paul. Ich war erstaunt, denn wir wohnten nur wenige Häuserblocks von einander entfernt. In diesem Telegramm teilte er mir mit, dass seine Freundin ihn aufgefordert hatte, sich innerhalb von 24 Stunden zwischen ihr und mir zu entscheiden, und dass er sich für sie entschieden hatte. Diese Entscheidung unterstützte er noch mit einem Zitat von Hegel, an dessen Wortlaut ich mich nicht mehr erinnern kann. Dann heirateten die beiden. Anscheinend hatte Paul dieser Frau immer erzählt, was er und ich in der vorigen Nacht besprochen hatten. Man kann ihr das Ultimatum nicht übel nehmen, denn Paul hatte ihr sicher auch gesagt, dass ich sie und ihre Freundin als Hyänen bezeichnet hatte. Jahre nach diesem Vorfall schrieb Feyerabend an Professor Hans Albert: „Den Kurt Fischer kenne ich sehr gut, wir waren sehr gute Freunde, haben uns aber eines verdammten Frauenzimmers wegen zerstritten (meine Schuld)."[18]

Unsere Freundschaft war beendet, und ich habe Feyerabend nur noch einmal kurz gesehen. Das war in Kassel, wo er an der Universität ein Seminar abhielt, zu dem auch Studenten aus Marburg angereist kamen. Ich war sehr beeindruckt, denn er fragte die Studenten, welche Probleme sie hätten, und es entwickelte sich eine interessante Diskussion. Am Ende des Seminars sagte Paul, dass sein Taxi schon warte und er daher sofort gehen müsse. Und er ging. Die Sache mit dem Taxi war mir sehr bekannt, und ich hätte mich gewundert, wenn kein Taxi auf ihn gewartet hätte. Später kam es noch zu einem indirekten Kontakt. Ein Verlag war gewillt, eine Festschrift zu meinem siebzigsten Geburtstag zu veröffentlichen aber nur unter der Bedingung, dass ein berühmter Philosoph unter den Herausgebern wäre. Dr. Cornelia Wegeler-Schardt, die zusammen mit Dr. Peter Muhr die Festschrift herausgeben wollte, wandte sich an Feyerabend, der zusagte. Hier einige Sätze aus seinem Zusagebrief: „Dass mein Name auf dem Buch Gelder zum Fließen gebracht hat, füllt mich mit Erstaunen. Sollte ich in meinem Alter noch respektabel werden? Ich hoffe nicht."[19]

Noch später luden mein Kollege Franz Wimmer und ich Feyerabend zu einem Symposion anlässlich des fünfzigsten Jahrestages des Anschlusses ein. Paul schickte mir eine Postkarte, auf der er meinte, dass Philosophie und Nationalsozialismus kaum etwas miteinander zu tun hätten und fügte noch hinzu: „Ich gehe die Sache anders an: Ich schreibe meine Autobiographie mit einem genauen Bericht über meine Gedanken und Handlungen während der Nazizeit."[20] Dieser Bericht fiel jedoch spärlich aus, meine ich. Ein Bezug auf die Nationalsozialisten fehlt in seiner Autobiographie, er findet sich höchstens in dem Satz, „Die Österreicher hatten Hitler mit überwältigtem Enthusiasmus begrüßt."[21] Ich war enttäuscht, denn von einem genauen Bericht kann wohl kaum die Rede sein. Er wusste wahrscheinlich nichts über die schrecklichen Grausamkeiten und Massenmorde und hätte ja auch nichts gegen sie unternehmen können, wenn er von ihnen gewusst hätte.

Nach dem Symposion schickte ich ihm den Tagungsband.[22] Er bedankte sich brieflich und schrieb, dass er nicht mehr in Berkeley sei, obwohl er hätte bleiben können. Ein Erdbeben habe ihn vertrieben. Das tue ihm leid, denn die Vereinigten Staaten von Amerika seien das einzige Land gewesen, das er irgendwie als Heimat empfunden habe.[23]

Ich komme zum Schluss. Paul Feyerabends wissenschaftliches Programm kann man noch zur Aufklärung zählen: es ist mit Fortschritt verbunden. Es handelt sich bei ihm um eine Strategie kleiner Fortschritte. Man könnte an das Motto der *Philosophischen Untersuchungen* Wittgensteins denken, das ein Zitat von Nestroy ist: „Überhaupt hat der Fortschritt das an sich, dass er größer ausschaut als er wirklich ist."[24] Und Nestroy war ja ein Lieblingsautor Feyerabends und Wittgenstein einer der Wenigen, die nicht von Feyerabend kritisiert wurden. Es gibt noch ein Indiz dafür, wie sehr er Nestroy schätzte. Über einen von Michael Benedikt und anderen herausgegebenen Band über die Philosophie in Österreich schreibt er in einem Brief an mich: „Leider, leider ist der Band nur bis 1820 – denn ich hätte gern gesehen, wie Nestroy dabei abschneidet."[25]

Den Wiener Kreis mit seinem Verifikationsprinzip hatte Feyerabend ebenso hinter sich gelassen wie den Falsifikationismus Poppers. Beide Prinzipien waren eigentlich große Visionen. Feyerabend setzt an ihre Stelle sein *Anything goes,* womit nicht nur ein neuer Anfang gemeint ist, eine neue Offenheit, sondern auch eine neue Schranke. In einem Gedankenaustausch mit Imre Lakatos wies er darauf hin, dass er die wissenschaftlichen Prinzipien, auf die sich sein Freund Lakatos stützte, ablehnte. *Anything goes* ist die Methodologie, die er in *Against Method* vertritt. *Against Method* wurde in drei Auflagen veröffentlicht und in neunzehn Sprachen übersetzt. In jeder Auflage gibt es Veränderungen und Zusätze. Aber hier sollen nicht Fragen nur skizziert werden, die in diesem Band an anderer Stelle ausführlicher und kompetenter behandelt werden. Nur so viel sei gesagt: Es wäre lohnend, einige der Konzepte Feyerabends genauer zu untersuchen, z.B. seinen Gebrauch des Begriffs Rationalismus. Bekannt ist dieser Begriff aus der Geschichte und der Philosophie. In der Geschichte wird er dem Irrationalismus gegenüber gestellt, dem Irrationalismus eines Nietzsche etwa. In der Philosophie wird er dem Empirismus gegenüber gestellt: Descartes, Leibniz und Spinoza sind Rationalisten, Locke, Berkeley und Hume sind Empiristen. Bei Feyerabend wird dieser Begriff anders verwendet, man könnte vielleicht sagen, phänomenologisch. Er schreibt:

> Die sozialen Gruppen, die vorbereiteten, was heute als der abendländische Rationalismus bekannt ist, und die die geistigen Grundlagen für die abendländische Wissenschaft legten, weigerten sich, diese Fülle so zu akzeptieren, wie sie sich darbot. Sie bestritten, dass die Welt so reichhaltig und das Wissen so komplex sei, wie es das handwerkliche Können und der Commonsense ihrer Zeit zu besagen schienen. Sie unterschieden zwischen einer „wirklichen Welt" und einer „Welt der Erscheinungen". Ihrer Ansicht nach war die wirkliche Welt einfach, gleichförmig, unveränderlichen allgemeinen Gesetzen unterworfen und für alle dieselbe. Neue Begriffe (die später „theoretische Begriffe" genannt wurden) waren nötig für ihre Beschreibung, und neue Fächer (die Erkenntnistheorie und, später, die Wisenschaftstheorie) entstanden bei dem Versuch zu erklären, wie sie zu erklären, wie sie zu dem Rest in Beziehung stand.[26]

Es sei noch angefügt, dass Feyerabend niemals „The Worst Enemy of Science" war, wie der Titel eines Aufsatzes im *Scientific American* behauptet.[27] Er war ein Gegner des Faches Wissenschaftstheorie, das er beim Deutschen Kongress für Philosophie als eine bisher unbekannte Form des Irrsinns bezeichnete.

Der letzte Satz seiner Autobiographie, kurz vor seinem Tod geschrieben lautet: „Das ist was ich mir wünsche: nicht, dass mein Geist weiterlebt, sondern allein die Liebe."[28] Ist Paul Feyerabend, nach einem Leben größter intellektueller Erfolge, auf eine Variation dessen zurückgekommen, was wohl in seinem Elternhaus gepredigt, wenn auch nicht praktiziert wurde: die Liebe? Sie spielt auch eine bedeutende Rolle in der Musik, eine seiner ganz großen Lieben. In einem Gespräch, das er mit Matthias Kroß 1992 kurz von seinem Tod geführt hat (zuerst veröffentlicht in der Berliner Zeitschrift *Zitty*), drückt Feyerabend seine Ansichten aus. Er begründet die geplante Abfassung seiner Autobiographie mit seinem Interesse für den Faschismus, von dem er meint, er hätte ihn „als Jugendlicher gar nicht verstanden." Und weiter sagt er: „Erst jetzt geht mir langsam auf, dass Humanität nicht über exakte Philosophie gelehrt werden kann." Vielleicht ist dies seine späte Einsicht, dass man zum Leben die Nähe zu den Menschen braucht, etwas, das Feyerabend schon viel mehr praktiziert hatte als er geglaubt hatte es getan zu haben.[29]

Anmerkungen

* Für Hilfe bei der Abfassung dieses Beitrags möchte ich mich bei Robert Kaller vom Institut ‚Wiener Kreis' bedanken.
1. Mit Feyerabend habe ich mich in „Paul Feyerabend: A Personal Reminiscence" in Kurt Rudolf Fischer, *Aufsätze zur angloamerikanischen und österreichischen Philosophie* (Frankfurt am Main–Berlin–Bern–New York–Paris–Wien 1999) beschäftigt und auch schon in Friedrich Stadler (Hrsg.), *Wissenschaft als Kultur. Österreichs Beitrag zur Moderne* (Wien–New York 1996).
2. Sigmund Freud, *Gesammelte Werke*, XV, S. 170.
3. Allerdings hatte auch der Logische Positivismus so etwas wie seine „übergeordnete Konstruktion", jedenfalls einen Satz, mittels dessen er ein Problem als prinzipiell lösbar oder unlösbar einordnen konnte: das Verifikationsprinzip.
4. Es braucht wohl nicht besonders hervorgehoben zu werden, dass der Antisemitismus im Ständestaat und im Dritten Reich völlig andere Auswirkungen hatte.
5. Siehe das 13. Kapitel in Kurt Rudolf Fischer, *Philosophie aus Wien* (Wien–Salzburg 1991) und auch den Beitrag in Friedrich Stadler (Hg.), *Vertriebene Vernunft*, II (Neuauflage, Münster 2004).
6. Philosophie war keine unabhängige, Erkenntnisse vermittelnde Disziplin sondern klärte nur die Grundsätze, Vorgangsweisen und Resultate anderer Disziplinen, also der Wissenschaften.
7. *Zeitverschwendung* (Frankfurt am Main 1995), S. 32f.
8. Ibid., S. 43f.
9. Ibid., S. 95.
10. Rudolf Haller, *Studien zur österreichischen Philosophie* (Amsterdam 1979), besonders Kapitel I, „Österreichische Philosophie".
11. Das Buch ist auch auf deutsch erschienen: *Sprache, Wahrheit und Logik* (Stuttgart 1970).
12. „Two Dogmas of Empiricism" ist zuerst in der *Philosophical Review* im Jänner 1951 erschienen.
13. Übersetzt als *Die Struktur wissenschaftlicher Revolutionen* (Frankfurt am Main 1967).
14. Dieser Kreis war auch als „Kraft-Kreis" bekannt.
15. Das 7. Kapitel von Hans Reichenbachs *Der Aufstieg der wissenschaftlichen Philosophie* (Berlin-Grunewald) ist eine Übersetzung von *The Rise of Scientific Philosophy* und trägt den Titel „Der Ursprung der neuen Philosophie".
16. Manche beschäftigen sich mit zwar relevanten, aber doch nicht eigentlich philosophischen Aufgaben wie z.B. der mathematischen Logik oder mit griechischen Texten.
17. Paul Feyerabend / Hans Albert, *Briefwechsel*. Hg. von Wilhelm Baum (Frankfurt am Main 1997)
18. Aus einem Brief von Feyerabend an Hans Albert vom 20.12.1969, in: *Briefwechsel*, S. 146.
19. Aus einem Brief Feyerabends an Cornelia Wegeler-Schardt vom 20.2.1992.
20. Aus einer Postkarte Feyerabends an Kurt Rudolf Fischer 16.6.1989.
21. *Zeitverschwendung*, S. 9.
22. Der Titel des Tagungsbandes ist *Philosophie und Politik an der Universität Wien 1930–1950* (Amsterdam 1992), herausgegeben von Kurt R. Fischer und Franz M. Wimmer.
23. Aus einem undatierten Brief Feyerabends an Kurt Rudolf Fischer.
24. Ludwig Wittgenstein *Philosophische Untersuchungen*. Siehe das Motto des Bandes.
25. S. Anmerkung 23.

26. *Irrwege der Vernunft* (Braunschweig 1979), S. 177.
27. John Horgan, "The Great Enemy of Science", *Scientific American*, May 1993.
28. *Zeitverschwendung*, S. 249.
29. Das Gespräch wurde 1992 in Zürich geführt. Hier bezeichnet Feyerabend die Nähe zu Menschen als wesentlich für das Leben und meint von der Philosophie, dass sie Distanz erzeugt (*Information Philosophie* 1/1995, S. 28-32).

Nachbemerkung: Die Schreiben Paul Feyerabends an Kurt Rudolf Fischer und Cornelia Wegeler-Schardt befinden sich im Institut Wiener Kreis.

PAUL HOYNINGEN-HUENE / ERIC OBERHEIM

NEUES ZU FEYERABEND

Einleitung

Der Titel dieses Aufsatzes, „Neues zu Feyerabend", wirft sogleich eine Frage auf: Für wen gibt es denn hier angeblich etwas Neues? Die Dinge, die wir diskutieren werden, waren zunächst einmal neu für E.O., der sie im Rahmen seiner Dissertation entdeckte. Sie waren dann auch neu für P.H.-H., der sie mit E.O. diskutieren und dann in dessen Dissertation lesen konnte, die dieser im Frühjahr 2004 an der Universität Hannover eingereicht hat. Die Dissertation trägt den Titel *On Feyerabend's Early Philosophy*. Neu sind die Dinge, die wir diskutieren werden, auch relativ zur Sekundärliteratur zu Feyerabend.[1] Einige der zu diskutierenden Dinge sind nicht mit dem Standardbild der Feyerabendschen Philosophie verträglich, wie es üblicherweise herumgeboten wird (meist als ein Schreckbild). Wir werden auf dieses (mittlerweile ziemlich langweilige) Standardbild allerdings nicht weiter eingehen, sondern gleich die Punkte nennen, die wir diskutieren werden. Zunächst wird es um zwei historische Neuheiten gehen. Erstens sind zwei bisher unbekannte Inspirationsquellen von Feyerabends Inkommensurabilitätsbegriff aufgetaucht, nämlich Pierre Duhem und Wolfgang Köhler.[2] Zweitens wird der Begriff von Realismus bei Feyerabend geklärt, der für seine Philosophie charakteristisch ist; darüber hat es kürzlich Kontroversen gegeben.[3] Dann werden wir vereinheitlichende konzeptionelle Aspekte von Feyerabends Philosophie diskutieren. Dazu gehört drittens das vielerorts in seiner Bedeutung für Feyerabend unterschätzte Mittel der immanenten Kritik. Viertens lässt sich eine philosophische Haltung identifizieren, die in vielerlei verschiedener Ausprägung ein zentraler Zielpunkt Feyerabendscher Kritik ist: der begriffliche Konservativismus. Positiv lässt sich fünftens Feyerabends Philosophie trotz allen Wandels, den sie natürlich auch aufweist, als eine pluralistische Erkenntnistheorie kennzeichnen. Und schließlich kann man sechstens noch eine weitere historische Frage stellen, die in der Literatur praktisch immer auf die selbstverständlichste Weise bejaht wird: War Feyerabend tatsächlich jemals ein Popperianer?

1. Inkommensurabilität

Also, was gibt es Neues zu den Inspirationsquellen von Feyerabends Inkommensurabilitätsbegriff? Zunächst scheint es ja, als sei Feyerabends Inkommensurabilitätsbegriff 1962 vom Himmel gefallen. In seinem berühmten Aufsatz Feyerabend (1962) besteht der Kern von Inkommensurabilität darin, dass wissenschaftliche Begriffe, insbesondere auch Beobachtungsbegriffe, wesentlich theorieabhängig sind.[4] Diese Idee hat Feyerabend aber nach eigenem Zeugnis schon viel früher entwickelt, nämlich in seiner Dissertation von 1951, und auch in weiteren Arbeiten vor 1962 verwendet – und diese Angaben Feyerabends sind korrekt (was bei Feyerabend nicht ganz selbstverständlich ist). Eine wesentlicher Referenzpunkt von Feyerabends Dissertation ist Pierre Duhems *Ziel und Struktur der physikalischen Theorien*, das dort öfters zitiert wird. Nun ist interessant, dass Duhem dem Inkommensurabilitätsbegriff zumindest sehr nahe ist, wie aus folgenden Zitaten deutlich wird:

> Das Resultat der Operationen, die den physikalischen Experimentator beschäftigen ist keineswegs eine Konstatierung einer Gruppe konkreter Tatsachen. Es ist der Ausdruck eines Urteils, das gewisse abstrakte symbolische Begriffe miteinander verbindet, deren Abhängigkeit von den wirklich beobachteten Tatsachen allein durch die Theorien hergestellt wird. [...] Die [...] Schlussfolgerungen [einer Abhandlung aus der Experimentalphysik] sind keineswegs die bloße einfache Darlegung gewisser Erscheinungen. Sie sind abstrakte Ausdrücke, denen wir keinen Sinn unterlegen können, wenn wir nicht die physikalischen Theorien kennen, auf die sich der Autor stützt. (Pierre Duhem, *Ziel und Struktur der physikalischen Theorien*. Hamburg: Meiner 1978 [1906], S. 193).

Hier ist die Vorstellung, dass es in der Wissenschaft theoriefreie Tatsachen gäbe, klar zurückgewiesen. Theorien dringen in das ein, was wie eine Tatsachenbeschreibung aussieht, und letztere ist nur verständlich, wenn man die Theorie kennt, die ihr zugrunde liegt.

> [D]as was der Physiker als Resultat eines Experiments ausspricht, ist nicht ein Bericht über konstatierte Tatsachen, ihre Versetzung in eine ideale, abstrakte, symbolische Welt, die durch die Theorie, die er als gültig betrachtet, geschaffen wurde. [...] Wenn die Theorien, die dieser Physiker als gültig betrachtet, dieselben sind, die wir an-

erkennen, [...] so sprechen wir in der gleichen Sprache und wir können uns verstehen. Aber es ist nicht immer so; es ist nicht so, wenn wir die Experimente eines Physikers diskutieren, der nicht unserer Schule angehört; es ist vor allem nicht so, wenn wir die Experimente eines Physikers diskutieren, den fünfzig Jahre, ein Jahrhundert, zwei Jahrhunderte von uns trennen. (*Ebd.*, S. 209-210).

Auch in diesem Zitat kommt einer der Kernpunkte der Inkommensurabilität zum Ausdruck, nämlich die Verständigungsschwierigkeiten, die sich zwischen inkommensurablen Standpunkten aufgrund der unterschiedlichen Theorien auftun, die für sie konstitutiv sind. Nirgends ist hier aber gesagt, genau so wenig wie bei Kuhn oder Feyerabend, dass diese Verständigungsschwierigkeiten prinzipiell unüberwindbar sind. Natürlich können wir lernen, die fremden Aussagen zu verstehen, wenn wir nur den theoretischen Hintergrund, der in sie eingegangen ist, aufgenommen haben. – Es ist also keine Frage, dass die Geschichte des Inkommensurabilitätsgedankens, so wie er relevant für die Wissenschaftsgeschichte ist, mindestens bis auf Duhem zurückgeht. Und es besteht ebenso kein Zweifel daran, dass Feyerabend mit Duhems einschlägigem Werk bei der Abfassung seiner Dissertation 1951 bestens vertraut war, wie sowohl aus seinem stark annotierten Exemplar von Duhems Buch und seinen Zitaten in der Dissertation hervorgeht.[5]

Aber es ist nicht nur der Gedanke der Inkommensurabilität, den man in Schriften findet, die Feyerabend früh bekannt waren, es ist sogar das Wort „Inkommensurabilität". Zunächst gibt es einen indirekten Einfluss Wolfgang Köhlers auf Feyerabend, nämlich über die *Philosophischen Untersuchungen* Wittgensteins, die Feyerabend 1952, vor ihrer Publikation, studiert hatte.[6] Er hatte ein Manuskript dieses Buches von Elizabeth Anscombe erhalten, die mit seiner Edition beschäftigt war.[7] In den *Philosophischen Untersuchungen* spielt die Gestaltpsychologie bekanntlich eine wichtige Rolle; Köhler ist einer der wenigen Autoren, die Wittgenstein namentlich nennt.[8] Es gibt nun einige starke Indizien dafür, dass Feyerabend die folgende Passage Köhlers vor 1962 gelesen hat.[9] Es handelt sich dabei um die 1938 erschienene englische Übersetzung eines Artikels von Wolfgang Köhler, der 1920 unter dem Titel „Die physischen Gestalten in Ruhe und im stationären Zustand" erschienen war:

In order to orient itself in the company of natural sciences, psychology must discover connections wherever it can between its own phenomena and those of the older disciplines. If this search fails, the psychology must recognize that its categories and those of natural science are incommensurable. (Wolfgang Köhler, „Physical Gestalten", in: W. Dennis (Hrsg.), *Readings in the History of Psychology.* New York: Appleton-Century-Crofts 1948, S. 513).

Das trifft auch der Sache nach präzise einen Aspekt von Feyerabends Inkommensurabilität, jedenfalls, wenn man die genannten „connections" als logische Relationen liest. Ebenfalls ist auffallend, dass Köhler als Relata der Inkommensurabilitätsrelation „Kategorien" ansetzt, genau wie Feyerabend, für den die Relata ebenfalls Begriffe sind (für den Kuhn von 1962 sind dagegen verschiedene Traditionen normaler Wissenschaft inkommensurabel). Es scheint daher keinen Zweifel zu geben, dass Feyerabend die Inspiration für seinen Begriff von Inkommensurabilität und sogar das Wort von früheren Autoren erhalten hat, insbesondere also von Köhler und Duhem.

2. Feyerabends Realismus

In der gegenwärtigen Diskussion um Feyerabend gibt es eine einflussreiche Interpretationstradition, die Feyerabend als einen wissenschaftlichen Realisten darstellt.[10] Wir hatten bereits in einer Arbeit von 1999 bestritten, dass das zumindest für den Feyerabend von 1962 zutrifft.[11] E.O. hat in der Folge diesen Problemkomplex in seiner Diskussion umfassend untersucht. Historisch kann man bei Feyerabend verschiedene Phasen seines Verhältnisses zum Realismus unterscheiden. Dabei sei aber sofort warnend angemerkt, dass die Zuschreibung von Positionen an Feyerabend immer problematisch ist und daher größte Vorsicht geboten ist, weil Feyerabend oft Positionen nur zeitweise und um des Argumentes willen annimmt und verteidigt (oder besser: anzunehmen und zu verteidigen scheint). Wir kommen auf dieses Problem im nächsten Abschnitt ausführlich zu sprechen.

Grob lassen sich drei Phasen in Feyerabends Stellungnahme zum Realismus unterscheiden. Von den 40er bis in die frühen 50er Jahre ist Feyerabend Positivist, d.h. in diesem Kontext, dass er die Frage nach dem Realismus bzw. dem Antirealismus als ein philosophisches Scheinproblem betrachtet. Diese Haltung erklärt sich zwanglos durch die philosophische Umgebung, in der er sich in dieser Zeit in Wien befindet.[12] In den frühen 50er Jahren wird er nun ein neokantischer Antirealist, d.h. er bezieht Position in dem Sinn, dass er einen unverstellten Zugang zu den Strukturen der Realität als unmöglich einschätzt, weil dieser Zugang notwendigerweise durch unsere Begriffe vermittelt ist.[13] Neokantisch ist seine Position, weil er diese realitätsvermittelnden Begriffe nicht als ein für allemal feststehend, sondern als historisch variabel ansieht. Zudem vertritt er eine weitere Position, die – für heutige Interpreten offenbar potentiell verwirrend – „normativer Realismus" heißt. Diese Position liegt überhaupt nicht in dem durch Realismus und Antirealismus aufgespannten Feld, ist also keine inhaltliche epistemologisch-ontologische Position. Vielmehr ist sie ein philosophisches Postulat an die empirischen Wissenschaften. Diese Forderung besagt, dass empirische Wissenschaften ihre Theorien realistisch interpretieren *sollen*. Diese Forderung sei für den Fortschritt der Wissenschaften ergiebiger als die konträre Forderung, die den Instrumentalismus favorisiert.[14] In den frühen 70er Jahren lässt Feyerabend den normativen Realismus fallen, weil er nun generell Forderungen an die Wissenschaften von Seiten der Philosophie ablehnt.[15] Die neokantische epistemologisch-ontologische Position behält er dagegen bei. Fatal ist nun die in der Literatur manchmal anzutreffende Verwechslung von Feyerabends *normativem* Realismus der mittleren Phase mit dem *wissenschaftlichen* Realismus,[16] weil sie Feyerabends wirkliche Position extrem verzerrt.

Nach diesen historischen Dingen kommen wir nun zu einigen inhaltlichen Aspekten von Feyerabends Philosophie. Generell erscheint sie, wenn man seine Schriften über einen bestimmten Zeitraum hin betrachtet, als etwas ziemlich heterogen, wenig konsistent, sprunghaft, anarchisch – anscheinend wie Feyerabend selbst.[17] Feyerabends Erklärungen dazu sind häufig auch nicht gerade vertrauenerweckend, wenn er sich gegen den Vorwurf, dass sich seine im Moment vertretene Position nicht mit einer kürzlich vertretenen decke, im wesentlichen dadurch verteidigt, dass er nicht mit dem Schmarrn von gestern belästigt werden wolle. Tatsächlich aber lassen sich in Feyerabends Philosophie methodische und inhaltliche Konstanten ausmachen, die seinem Werk viel größere Einheitlichkeit und Geschlossenheit verleihen als das bisher angenommen worden war. Methodisch ist das vor allem die Figur der immanenten Kritik (Abschnitt 3), inhaltlich seine Frontstellung gegen den begrifflichen Konservativismus (Abschnitt 4) und seine Verteidigung einer pluralistischen Erkenntnistheorie (Abschnitt 5).

3. Immanente Kritik

Es ist charakteristisch für Feyerabend bis in die 70er oder 80er Jahre, dass er die meisten seiner Ideen in der kritischen Auseinandersetzung mit Anderen entwickelt – er ist mit vielen im (z.T. sehr polemischen) Gespräch. Als Kritikstrategie verwendet er fast ausschließlich die der immanenten Kritik. Dabei wird die zu kritisierende Position nicht mit einem alternativen Standpunkt und zugehörigen Argumenten konfrontiert, sondern man versucht eine Zersetzung der zu kritisierenden Position gewissermaßen von innen. Dabei werden keinerlei der zu kritisierenden Position nicht zugehörigen Annahmen gemacht, vielmehr wird versucht, uneingestandene Voraussetzungen und Annahmen der Position oder Konsequenzen von ihr aufzudecken, die für ihre Verteidiger unakzeptierbar sind. Oberflächlich betrachtet sieht eine solche Strategie so aus, als würde der Kritiker die Position selbst annehmen und vertreten. Tatsächlich aber stellt er sich nur um des Argumentes willen auf ihren Standpunkt, nämlich um ihre internen Probleme ans Licht zu bringen. Sehr viele von Feyerabends Arbeiten haben diese Struktur, was in der Literatur aber immer wieder übersehen wird.[18] Aus solchen Auseinandersetzungen mit kritisierten Positionen lässt sich *niemals* positiv auf die Position Feyerabends schließen, denn Prämissen, die er in seiner Argumentation benutzt, sind nicht seine Prämissen, sondern die der Gegenposition. Welche Position Feyerabend selbst in der fraglichen Sache vertritt, bleibt völlig im Hintergrund, und vielfach kann es auch sein, dass sich Feyerabend dazu noch gar keine Meinung gebildet hat oder gar keine Meinung bilden will. Das Verfahren der immanenten Kritik ist Feyerabends *primärer Weg der philosophischen Auseinandersetzung*, weil sich dadurch eine Klärung von vorgegebenen und eventuell Hinweise auf alternative Positionen ergeben; dieser Weg ist also durchaus nicht rein negativ. Die Ähnlichkeit mit Hegel, der in der Geschichte der Philosophie die immanente Kritik wohl am konsequentesten als fundamentales philosophisches Arbeitsmittel eingesetzt hat, ist hier unübersehbar. Interessanterweise hat sich Feyerabend aber erst in den späten 1960er Jahren mit Hegel auseinandergesetzt,[19] dann aber vor allem mit der Logik, nicht aber mit der Phänomenologie des Geistes.[20]

Ein nicht unerheblicher Teil der Verwirrung um Feyerabends Philosophie rührt nun daher, dass der große Anteil von immanenter Kritik in seinen Schriften nicht wahrgenommen wird. Stattdessen werden ihm die kritisierten Positionen zugeschrieben, was zu dem Eindruck führt, dass Feyerabend seine eigene Position nach Lust und Laune wechselt, ohne dass gute Gründe dafür oder eine sonstige irgendwie geartete Linie sichtbar würde. Kurz: In dieser verzerrten Perspektive erscheint Feyerabend zwar als ein scharfsinnig argumentierender, aber insgesamt unseriöser, weil sophistischer Philosoph.

4. Gegen den begrifflichen Konservativismus

Eines der mächtigsten und dauerhaftesten Motive in Feyerabends Philosophieren ist seine Opposition gegen den begrifflichen Konservativismus. Mit „begrifflichem Konservativismus" ist die Tendenz gemeint, in bestimmten historischen Kontexten entstandene Begriffe der Wissenschaften und der Philosophie für unwandelbar und endgültig zu halten. Diese Tendenz findet sich in sehr unterschiedlichen Gebieten. Wohl aufgrund dieser Heterogenität ist Feyerabends grundlegende Kritik am begrifflichen Konservativismus in seiner enorm vereinheitlichen Kraft für sein Philosophieren bislang nicht gesehen worden. Viele anscheinend zusammenhangslose kritische Einzelanalysen Feyerabends, deren Heterogenität man bisher nur achselzuckend zur Kenntnis nehmen und allenfalls mit psychologischen Erklärungen garnieren konnte, werden dadurch verbunden. Das Bild eines sprunghaften, chaotischen und launischen Denkers tritt zugunsten eines konsequenten philosophischen Kritikers begrifflicher Verhärtungen zurück.[21] Zudem besteht eine enge Verbindung von Feyerabends Kritik am begrifflichen Konservativismus mit seiner Inkommensurabilitätsthese, was in seinen Arbeiten nur punktuell, nicht aber in Allgemeinheit zum Ausdruck kommt. Inkommensurabilität ist eine unmittelbare Konsequenz des Aufgebens begrifflich konservativer Verhältnisse, wo dieses nämlich zu einer Veränderung von Begriffen führt, die dann mit den ursprünglichen Begriffen inkommensurabel sind.

Wir können einige Beispiele für Feyerabends Kritik am begrifflichen Konservativismus hier nur nennen, nicht aber detailliert diskutieren.[22] Gemeint ist Feyerabends Kritik an
- der Transzendentalphilosophie[23]
- Bohrs Komplementaritätsthese bezüglich klassischer und Quantenphysik[24]
- Heisenbergs Konzept abgeschlossener Theorien[25]
- Wittgensteins später Philosophie[26]
- Nagels Theorie der Reduktion[27]
- Kuhns normaler Wissenschaft[28]

– der Idee, Wissenschaftsentwicklung als Annäherung an die Wahrheit zu sehen (Peirce, Popper)[29]

In allen diesen heterogenen Gebieten besteht Feyerabends Kritik darin, dass er die Zementierung bestimmter begrifflicher Verhältnisse als für den Erkenntnisfortschritt hinderlich ansieht. Es ist die begriffliche Offenheit, die Feyerabend in diesen Arbeiten für die Wissenschaften und die Philosophie propagiert. Sie ist für den Aufklärer Feyerabend charakteristisch.

5. Pluralistische Erkenntnistheorie

Wenn man Feyerabend überhaupt eine Erkenntnistheorie zuschreiben kann, dann ist es eine pluralistische Erkenntnistheorie.[30] Eine solche pluralistische Erkenntnistheorie hat zwei Aspekte. Auf der Objektebene der wissenschaftlichen Erkenntnis fordert eine solche Theorie, dass in den Wissenschaften ein Pluralismus im Sinne der Entwicklung von Alternativen zu bestehenden Theorien notwendig sei. Auf der Metaebene der philosophischen Reflexion fordert eine solche Theorie, dass hier ein Pluralismus im Sinne verschiedener philosophischer Zugangsweisen zu ein und demselben Objektbereich notwendig sei. Der Kern des Arguments für diese beiden Aspekte einer pluralistischen Erkenntnistheorie ist, dass nur ein solcher Pluralismus wirklich erkenntnisförderlich sei; wir werden auf dieses Argument gleich detaillierter eingehen. Die Alternative zu diesem Pluralismus ist in Feyerabends Sicht die wissenschaftliche bzw. philosophische dogmatische Erstarrung.[31]

Die Überzeugung von der Notwendigkeit einer pluralistischen Erkenntnistheorie zieht sich durch Feyerabends gesamtes philosophisches Schaffen seit 1947, zugegebenermaßen in verschiedenen Graden der expliziten Artikulation. Wird man sich dieses Umstands bewusst, so wird ein weiteres enorm vereinheitlichendes Motiv in Feyerabends philosophischem Werk sichtbar. Worin bestehen nun Feyerabends Argumente für eine solche pluralistische Erkenntnistheorie? Wir werden nun die beiden genannten Teilaspekte dieser Theorie unterscheiden müssen, und zunächst die Argumente für den wissenschaftlichen Theorienpluralismus und dann die Argumente für den Pluralismus philosophischer Zugangsweisen diskutieren.

Feyerabends Argumente für den Pluralismus auf der Objektebene sind in der Literatur verhältnismäßig gut bekannt, und darum können wir uns kurz fassen. Die empirische Überprüfung einer Theorie soll typischerweise nicht in einer simplen Konfrontation dieser einen Theorie mit empirischen Daten bestehen, sondern in einer Theorienkonkurrenz mehrerer Theorien.[32] Zu entscheiden ist dabei die Frage, welche von ihnen die größte Übereinstimmung mit den empirischen Daten aufweist. Der Test einer bestimmten Theorie wird durch dieses Verfahren generell strenger, was der Grund für den Vorzug der Theorienkonkurrenz vor der simplen Konfrontation einer Theorie mit Daten ist. Darüber hinaus besteht nach Feyerabend sogar manchmal, insbesondere bei fundamentalen Theorien, die *Notwendigkeit* einer solchen Theorienkonkurrenz, soll es überhaupt zu einem wirklichen Test der fraglichen Theorie kommen.[33] Der Grund ist, dass vielfach nur im Lichte konkurrierender Theorien entschieden werden kann, ob es sich bei einem Phänomen, das für die Theorie eine Anomalie darstellt, um eine seriöse Abweichung von der Theorie und damit um eine Falsifikation oder um einen bloßen „Dreckeffekt" handelt. In anderen Worten, die Beurteilung, ob ein bestimmtes anomales Phänomen als Dreckeffekt beiseite geschoben werden kann oder als falsizifierende Instanz ernst genommen werden muss, kann oft nur vor dem Hintergrund einer alternativen Theorie gefällt werden.[34] Daher sind für die kritische Konfrontation von Theorien mit der Empirie Alternativtheorien erwünscht (zur Verschärfung des Tests) und z.T. sogar notwendig (zur Konstitution eines Tests). Daher also Feyerabends Aufruf zur „Theorienproliferation".

Was in der Literatur aber vollkommen unbekannt zu sein scheint ist der biographische und sachliche Hintergrund, der Feyerabend zur Ausarbeitung dieser Position geführt hat.[35] Feyerabend wurde 1947 in Wien mit dem unorthodoxen Experimentalphysiker Felix Ehrenhaft konfrontiert. Dieser produzierte in verschiedenen Bereichen der Physik experimentelle Ergebnisse, die nicht mit den akzeptierten Theorien in Einklang standen. Hätte man diese Ergebnisse gewissermaßen zum Nennwert genommen, so hätte man einen großen Teil der akzeptierten Theorien der Physik verwerfen müssen – ein Ergebnis, das für den wissenschaftsbegeisterten Feyerabend von 1947 absolut unakzeptierbar war. Natürlich genügte es umgekehrt auch nicht, die abweichenden Daten allein von der in Frage stehenden wohletablierten Theorie aus zu beurteilen und damit als offensichtlich abwegig zu verwerfen. Vielmehr musste eine andere, außerhalb stehende Beurteilungsinstanz beigezogen werden, um das Gewicht der abweichenden Daten zu beurteilen, wenn eine interne Analyse der Datengewinnung keine Fehler zutage brachte. Nachdem Ehrenhaft bei der Gewinnung seiner Daten keine Fehler nachgewiesen werden konnten, mussten also von der zu testenden Theorie unabhängige Argumente aufgeboten werden, wenn man den potentiell falsifizierenden Charakter seiner Daten nicht dogmatisch leugnen, sondern argumentativ neutralisieren wollte. Und genau dazu waren andere Theorien beizuziehen.

Feyerabends Argumente für den Pluralismus auf der Metaebene haben ihre Wurzel in Denkerfahrungen, die man in der Philosophie selbst machen kann. Zum einen führt der spekulative Charakter philosophischer Thesen dazu, dass sie nicht wirklich abschließend begründbar sind, selbst wenn sie sich in einer bestimmten Epoche sozial mehr oder weniger vollständig durchsetzen konnten. Immer wieder sind in der Geschichte der Philosophie anscheinend endgültig gewonnene philosophische Einsichten von späteren Generationen wieder in Frage gestellt worden. Zum anderen zeigt sich in der Philosophie immer wieder, dass verschiedene, auch konträre philosophische und sogar wissenschaftliche Zugänge zu einem komplexen Problem verschiedene Facetten des Problems erschließen. Deswegen ist niemals auszuschließen, dass auch ein Zugang, der im Moment aus welchen Gründen auch immer nicht en vogue ist, für die Problemlösung relevant sein kann. Philosophie sollte daher ein Unternehmen sein, das sich durch größtmögliche methodische Offenheit auszeichnet.

Es ist interessant, dass Feyerabend diesen Pluralismus lange vor *Against Method* (1975) praktiziert, aber erst Anfang der 70er Jahre explizit artikuliert und mit *Against Method* weit herum bekannt gemacht hat. Es könnte dies ein Hinweis darauf sein, dass es auch in der Philosophie den Fall gibt, dass eine bestimmte Zugangsweise praktiziert wird, ohne dass das ihren Protagonisten gänzlich bewusst und durchsichtig ist. Erst später, nach Reflexion auf die eigene Praxis, können die Akteure ein mehr oder weniger vollständiges Bewusstsein über ihre eigene Tätigkeit gewinnen.

6. War Feyerabend jemals ein Popperianer?

In der Literatur findet man die folgende weit verbreitete Meinung über Feyerabends philosophische Entwicklung. Feyerabend wurde nach seiner anfänglichen Beschäftigung mit dem logischen Positivismus in Wien in den 50er Jahren ein Popperianer, radikalisierte diese Philosophie in den 60er Jahren und wandte sich schließlich in den 70er Jahren mit großem Getöse von ihr ab. Um diese Einschätzung zu überprüfen, insbesondere ob Feyerabend in den 50er Jahren ein Popperianer war, sollte man zunächst analytisch zwei Aspekte der hier implizit vorkommenden Relation „x ist Mitglied der Schule y" unterscheiden. Einmal gibt es einen soziologischen Aspekt von „x ist Mitglied der Schule y". Im soziologischen Sinn ist man Mitglied einer Schule, wenn ein informelles Analogon zur formellen Mitgliedschaft in einer Partei oder einem Verein besteht. Eine solche informelle Mitgliedschaft kann sich auf verschiedene Weisen äußern, z.b. durch Selbst- oder Fremddeklaration, durch Anwesenheit bei Treffen der Schule, durch Kommunikationsnetzwerke und vieles mehr. Im epistemologischen Sinn ist man Mitglied einer Schule, wenn man Anhänger einer Mehrzahl der wesentlichen Thesen der Schule ist. – Zugegebenermaßen ist diese Charakterisierung der zwei Bedeutungen von Schulzugehörigkeit vorläufig und vage, aber sie macht auf verschiedene Aspekte von Schulzugehörigkeiten aufmerksam. Natürlich gehen diese beiden Aspekte oft und typischerweise Hand in Hand, aber es besteht kein zwingendes Implikationsverhältnis zwischen beiden Aspekten von Schulzugehörigkeit. So kann man aus sozialen Gründen die Nähe einer bestimmten philosophischen Schule suchen, ohne wesentliche ihrer Thesen zu teilen, und man kann Grundthesen einer Schule teilen, ohne dass man sich ihr in einem sozialen Sinn anschließt (ganz ähnlich, wie das für das Verhältnis zu Religionsgemeinschaften gilt). Es empfiehlt sich daher, die beiden Aspekte der Schulzugehörigkeit bei Feyerabend hinsichtlich der Popper-Schule separat voneinander zu prüfen.

War Feyerabend in einem soziologischen Sinn Mitglied der Popper-Schule? Dem steht zunächst entgegen, dass Feyerabend zumindest später bestritten hat, je Mitglied der Popper-Schule gewesen zu sein.[36] Allerdings sind Feyerabends Vergangenheitsbeschreibungen mit größter Vorsicht zu genießen, wenn Popper in irgendeiner Weise mit im Spiel ist. Denn ganz offensichtlich hat sich sein Verhältnis zu Popper im Laufe der Jahre massiv verändert, und Feyerabend hat – aus welchen Gründen auch immer – aktiv dazu beigetragen, die Art seines frühen Verhältnisses zu Popper zu verschleiern. So gibt es in Feyerabends Schriften bis in die Mitte der 60er Jahre erhebliche Lobeshymnen auf Popper, von denen aber viele beim Wiederabdruck der entsprechenden Arbeiten durch Feyerabend eliminiert bzw. verändert wurden. So gibt es z.B. im Wiederabdruck einer Arbeit Feyerabends eine Passage, in der Duhem sehr positiv erwähnt wird, wo sich aber im Original an der gleichen Stelle statt „Duhem" der Name „Popper" findet.[37] – Aufgrund dieser schwierigen Datenlage schieben wir die Beantwortung der Frage nach Feyerabends sozialer Zugehörigkeit zur Popper-Schule etwas auf.

Aber war Feyerabend nicht in den 50er Jahren ein Anhänger der Popper-Schule im epistemologischen Sinn, d.h. hat er nicht viele der zentralen Lehrstücke der Popper-Schule vertreten? Richtig ist, dass PKF in den 50er Jahren viele Elemente des kritischen Rationalismus positiv würdigt und auch selbst verwendet. Dazu gehört beispielsweise die These, dass Nachfolgertheorien typischerweise im logischen *Widerspruch* zu ihren Vorgängertheorien stehen, so dass sie keineswegs in einem einfachen Sinn als Verallgemeinerungen ihrer Vorgängertheorien aufgefasst werden können. Aus dieser Einsicht ergibt sich eine wesentliche Kritik am Induktivismus, bei dem die induktive Verallgemeinerung von Daten oder spezialisierten Theorien eine tragende Rolle spielt. Das gibt seinerseits das Motiv für den Falsifikationismus und den Kritizismus ab, die schließlich als zentrale Doktrinen für die Popper-Schule charakteristisch sind. Ebenfalls teilt Feyerabend mit der Popper-Schule das Konzept einer normativen Wissenschaftstheorie.

Allerdings gibt es auch Argumente, die dagegen sprechen, dass Feyerabend im epistemologischen Sinn ein Popperianer war. Erstens ist die häufige Verwendung von Thesen des kritischen Rationalismus nicht nur mit der Mitgliedschaft bei der Popper-Schule verträglich, sondern auch damit, dass Feyerabend ein Pluralist ist, der alle möglichen Philosophien verwendet, darunter auch den kritischen Rationalismus (wenn auch am häufigsten). Daher kann aus Feyerabends häufiger Diskussion und Verwendung von Thesen aus dem kritischen Rationalismus eine Mitgliedschaft bei dieser Schule *nicht abgeleitet* werden. Zweitens entfernt sich Feyerabend mit seiner Inkommensurabilitätsthese an einem zentralen Punkt vom kritischen Rationalismus. Er geht nämlich über dessen These, dass eine Nachfolgertheorie typischerweise der ursprünglichen Theorie *widerspricht* und sie *korrigiert*, für den Fall universeller Theorien weit hinaus.[38] Für diese Theorien behauptet Feyerabend ja, dass die Nachfolgertheorie *keine logischen Beziehungen* zur früheren hat und sie nicht korrigiert, sondern *ersetzt*. Dies hat drittens eine Konsequenz, die wieder einer zentralen Doktrin der Popper-Schule widerspricht, dass man nämlich die Abfolge fundamentaler Theorien *nicht* als eine Annäherung an die Wahrheit betrachten kann.

In epistemologischer Hinsicht ist also die Frage nach der Zugehörigkeit Feyerabends zur Popper-Schule längst nicht so klar, wie es zunächst scheint. In den letzten beiden Punkten steht Feyerabend tatsächlich Thomas Kuhn viel näher als der Popper-Schule, und unter anderem diese letzten beiden Punkte haben Kuhn den entschiedenen Widerspruch von Seiten der Popperianer eingebracht.

Dieser Zweifel an der Zugehörigkeit zur Popper-Schule im epistemologischen Sinn wirft auch Zweifel an seiner Zugehörigkeit zur dieser

Schule im soziologischen Sinn auf (wie oben ausgeführt, ist die Frage der Schulzugehörigkeit im soziologischen Sinn damit aber *nicht entschieden*). Aber wenn man annimmt, dass Feyerabend wenigstens im soziologischen Sinn Mitglied der Popper-Schule war, dann bedürfte ein weiteres irritierendes Faktum einer Erklärung. Warum hat Feyerabend im Jahre 1952 das Angebot Poppers sein Assistent zu werden, ohne das Vorhandensein von beruflichen Alternativen ausgeschlagen (die für Feyerabend nicht abwegige Erklärungsheuristik „cherchez la femme" scheint hier ergebnislos zu sein)? Wäre es für ein junges Mitglied einer Schule nicht ein überaus attraktives, ja irresistibles Angebot gewesen, an einer berühmten Hochschule Assistent des unumschränkten Schulhaupts zu werden?

7. Resultat

Spätestens aufgrund dieser neuen Resultate kann man sagen, dass die Philosophie Feyerabends mit ihren realen Stärken und Schwächen und in ihrer Entwicklung noch nicht gut verstanden ist. Dies gilt keineswegs nur für die deutschsprachige Literatur, sondern ebenso für die (international dominierende) englischsprachige. Wie das bei der Rezeption der Philosophie Kuhns der Fall war, so ist auch die Rezeption Feyerabends von vielen Oberflächlichkeiten und Klischees durchsetzt. Dies gilt nicht nur für die an Wissenschaftstheorie interessierten Fachwissenschaften und das Feuilleton, sondern auch für die Philosophie selbst. Dies ist bedauerlich, denn allem Anschein nach ist das Anregungspotential dieses nicht nur durch seine Farbigkeit auffallenden Philosophen noch nicht ausgeschöpft.

Anmerkungen

1. Das gilt insbesondere für John Preston, *Feyerabend. Philosophy, Science and Society.* Oxford: Blackwell 1997.
2. Siehe Eric Oberheim, „On the Historical Origins of the Contemporary Notion of Incommensurability: Paul Feyerabend's Assault on Conceptual Conservativism", in: *Studies in the History and Philosophy of Science.* Im Druck.
3. Siehe John Preston, „Review Symposia: Radical Fallibilism vs. Conceptual Analysis: The significance of Feyerabend's Philosophy of Science. Author's Response", in: *Metascience* 8, 2, 1998, S. 233-243; Howard Sankey, „Incommensurability: The Cur-

rent State of Play", in: *Theoria* 12, 1997, S. 425-445; und Eric Oberheim und Paul Hoyningen-Huene, „Review Symposium: Radical Fallibilism vs. Conceptual Analysis: The Significance of Feyerabend's Philosophy of Science", in: *Metascience. The Australian Journal for The Philosophy of Science* 8, 2, 1998, S. 226-233.
4. Paul Feyerabend, „Explanation, Reduction and Empiricism", In: Herbert Feigl und Grover Maxwell (Hrsg.), *Scientific Explanation, Space, and Time. Minnesota Studies in the Philosophy of Science, Vol. III.* Minneapolis: University of Minnesota Press, 1962, S. 28-97.
5. Feyerabends annotiertes Exemplar von Duhems Buch ist im philosophischen Archiv der Universität Konstanz zu finden. Darin sind die oben zitierten Passagen hervorgehoben. Es gibt gute Gründe anzunehmen, dass diese Anmerkungen um 1950 gemacht worden sind.
6. Ludwig Wittgenstein, *Philosophical Investigations.* Oxford: Basil Blackwell 1953.
7. Siehe Paul Feyerabend, *Killing Time. The Autobiography of Paul Feyerabend.* Chicago: University of Chicago Press 1995, S. 92-93 und Paul Feyerabend, *Science in a Free Society.* London: New Left Books 1978, S. 115.
8. Siehe Ludwig Wittgenstein, *Philosophical Investigations*, a.a.O., S. 203^e.
9. Der Terminus 'incommensurable' wurde von Feyerabend erstmals benutzt, um Nagels Sichtweise der formalen Bedingungen der Reduktion zu kritisieren. Nagel entwickelte diese Sichtweise in Kapitel 11, „The reduction of theories", von Ernst Nagel, *The Structure of Science. Problems in the Logic of Scientific Explanation.* London: Routledge 1966 [1961]. Hier führten Nagels Betrachtungen einiger potenzieller Probleme seiner Reduktionstheorie zu Wolfgang Köhler, „Physical Gestalten", a.a.O, siehe Ernst Nagel, *The Structure of Science. Problems in the Logic of Scientific Explanation.* a.a.O., S. 392, n. 22 und n. 24, und S. 393, n. 26. Genau darin verwendet Köhler den Terminus „incommensurable", um die Beziehung zwischen einigen Begriffen der Psychologie und der Physik zu beschreiben. Dies ähnelt einem von Feyerabend häufig verwendeten Beispiel von Inkommensurabilität, siehe Paul Feyerabend, „Explanation, Reduction and Empiricism", a.a.O.; Paul Feyerabend, „Materialism and the Mind-Body Problem", in: *The Review of Metaphysics* 17, 1963, S. 49-66; und Paul Feyerabend, „Mental Events and the Brain", in: *The Journal of Philosophy* 60, 1963, S. 295-296. Es erscheint unwahrscheinlich, dass es sich hierbei um einen reinen Zufall handelt. Eher drängt sich der Eindruck auf, dass Feyerabend zur Kritik an Nagel dessen Hauptquellen heranzog, welche von Nagel bereits mit Bezug auf die Probleme der Reduktion erwähnt worden waren. In diesem Zusammenhang stieß Feyerabend auf den Terminus „incommensurable".
10. Siehe z.B.. Howard Sankey, *The Incommensurability Thesis.* Aldershot: Avebury 1997; Howard Sankey, „Incommensurability: The Current State of Play", a.a.O.; John Preston, *Feyerabend. Philosophy, Science and Society.* a.a.O.; und John Preston, „Feyerabend's Retreat from Realism", *Philosophy of Science* 64, 1997, S. 421-431.
11. Eric Oberheim und Paul Hoyningen-Huene, „Review Symposium: Radical Fallibilism vs. Conceptual Analysis: The Significance of Feyerabend's Philosophy of Science", a.a.O.
12. See Paul Feyerabend, *Killing Time. The Autobiography of Paul Feyerabend*, a.a.O., S. 72. Hollitscher überzeugte Feyerabend davon, die Position eines strengen Positivismus einzunehmen, welcher von Popper und Kraft im Rahmen des normativen Realismus formuliert wurde.
13. Siehe Paul Feyerabend, „Explanation, Reduction and Empiricism", a.a.O., S. 29, sowie Paul Feyerabend, „Review of Kant's Metaphysics and Theory of Science. By Gottfried Martin", in: *British Journal for the Philosophy of Science* 7, 1956, S. 260-263; Paul Feyerabend, „An Attempt at a Realistic Interpretation of Experience", in:

Proceedings of the Aristotelian Society 58, 1958, S. 143-170, hier S. 150-151, 158; Paul Feyerabend, „Complementarity", in: *Proceedings of the Aristotelian Society, 58, Suppl. Vol. 32,* 1958, 75-104, hier S. 81-84; Paul Feyerabend, „Review of Erkenntnislehre. By V. Kraft", in: *British Journal for the Philosophy of Science* 13, 1963, 319-323, hier S. 319 und 322; Paul Feyerabend, „Reply to Criticism. Comments on Smart, Sellars and Putnam", in: Robert Cohen und Mark Wartofsky (Hrsg.), *Proceedings of the Boston Colloquium for the Philosophy of Science 1962–64: In Honor of Philipp Frank. Boston Studies in the Philosophy of Science, Vol. II.* New York: Humanities Press 1965, S. 223-261, hier S. 246-247; Paul Feyerabend, „Problems of Empiricism", in: R. Colodny (Hrsg.), *Beyond the Edge of Certainty. Essays in Contemporary Science and Philosophy.* Pittsburgh: CPS Publications in the Philosophy of Science 1965, S. 145-260, hier S. 216, 221, n. 8 und 259, n. 161; und Paul Feyerabend, *Science in a Free Society,* a.a.O., hier S. 171, n. 38.
14. Siehe Paul Feyerabend, „Realism and Instrumentalism: Comments on the Logic of Factual Support", in: Mario Bunge (Hrsg.), *The Critical Approach to Science and Philosophy: In Honor of Karl R. Popper.* London: The Free Press of Glencoe 1964, S. 280-308.
15. See Paul Feyerabend, *Science in a Free Society.* a.a.O., S. 117, 144, n. 7, und 148.
16. Siehe z.B. John Preston, *Feyerabend. Philosophy, Science and Society.* a.a.O.; und John Preston, "Feyerabend's Retreat from Realism", a.a.O.
17. Zu Feyerabend als Person siehe seine Autobiographie *Killing Time. The Autobiography of Paul Feyerabend,* a.a.O.; sowie Paul Hoyningen-Huene, „Paul K. Feyerabend", in: Journal for General Philosophy of Science 28, 1, 1997, S. 1-18.
18. Siehe z.B. Paul Feyerabend, „An Attempt at a Realistic Interpretation of Experience", a.a.O.; Paul Feyerabend, „Complementarity", a.a.O.; Paul Feyerabend, „On the Interpretation of Scientific Theories", in: *Proceedings of the 12th International Congress of Philosophy, Venice, 1958, Vol. 5.* Florence: Sansoni Editore 1960, S. 151-159; Paul Feyerabend, „Explanation, Reduction and Empiricism", a.a.O.; Paul Feyerabend, „How to Be a Good Empiricist: A Plea for Tolerance in Matters Epistemological", in: B. Baumrin (Hrsg.), *Philosophy of Science: The Delaware Seminar.* New York: Interscience Press 1963, S. 3-39; und Paul Feyerabend, „Problems of Empiricism", a.a.O.
19. Siehe Wilhelm Baum (Hrsg.), *Paul Feyerabend – Hans Albert. Briefwechsel.* Frankfurt am Main: Fischer 1997.
20. Letztere hat er nach eigener Auskunft nie gelesen (persönliche Mitteilung an P.H.-H. in den 80er Jahren).
21. Wieder ist eine Ähnlichkeit mit Hegel unübersehbar, wenn auch bei Feyerabend die Kritik der Begriffe keinerlei einheitliche Zielorientierung hat.
22. Siehe dazu Eric Oberheim, „On the Historical Origins of the Contemporary Notion of Incommensurability: Paul Feyerabend's Assault on Conceptual Conservativism", a.a.O.
23. Feyerabend hat immer wieder die Idee der transzendentalen Notwendigkeit zurückgewiesen, siehe z.B. Paul Feyerabend, „Explanation, Reduction and Empiricism", a.a.O., S. 69-70; Paul Feyerabend, „Problems of Empiricism", a.a.O., S. 247; und Paul Feyerabend, „Reply to Criticism. Comments on Smart, Sellars and Putnam", a.a.O., S. 244.
24. Siehe z.B. Paul Feyerabend, „An Attempt at a Realistic Interpretation of Experience", a.a.O.; Paul Feyerabend, „Niels Bohr's Interpretation of the Quantum Theory", in: Herbert Feigl und Grover Maxwell (Hrgs.), *Current Issues in the Philosophy of Science. Symposia of Scientists and Philosophers. Proceedings of Section L of the American Association for the Advancement of Science, 1959.* New York: Holt,

Rinehart and Winston 1961, S. 371-390; Paul Feyerabend, „Explanation, Reduction and Empiricism", a.a.O.; Paul Feyerabend, „Problems of Microphysics", in: R. Colodny (Hrsg.), *Frontiers of Science and Philosophy: University of Pittsburgh Series in the Philosophy of Science Vol. I*. Pittsburgh: University of Pittsburgh Press 1962, S. 189-283; Paul Feyerabend, „How to Be a Good Empiricist: A Plea for Tolerance in Matters Epistemological", a.a.O.; Paul Feyerabend, „Problems of Empiricism", a.a.O.; und Paul Feyerabend, „Against Method: Outline of an Anarchistic Theory of Knowledge", in: M. Radner und S. Winokur (Hrsg.), *Analysis of Theories and Methods of Physics and Psychology. Minnesota Studies in the Philosophy of Science vol. IV*. Minneapolis: University of Minnesota Press 1970, S. 17-130.
25. Siehe Paul Feyerabend, *Zur Theorie der Basissätze*. Universitätsbibliothek Wien 1951, sowie Eric Oberheim, „On the Historical Origins of the Contemporary Notion of Incommensurability: Paul Feyerabend's Assault on Conceptual Conservativism", a.a.O.
26. Siehe z.B. Paul Feyerabend, „Explanation, Reduction and Empiricism", a.a.O.; Paul Feyerabend, „How to Be a Good Empiricist: A Plea for Tolerance in Matters Epistemological", a.a.O.; Paul Feyerabend, „Problems of Empiricism", a.a.O.; Paul Feyerabend, „Linguistic Arguments and Scientific Method", in: *Telos* 2, 1969, S. 43-63; und Paul Feyerabend, „Against Method: Outline of an Anarchistic Theory of Knowledge", a.a.O.
27. Siehe z.B. Paul Feyerabend, „Explanation, Reduction and Empiricism", a.a.O.; und Paul Feyerabend, „How to Be a Good Empiricist: A Plea for Tolerance in Matters Epistemological", a.a.O.
28. Siehe z.B. Paul Feyerabend, „Two Letters of Paul Feyerabend to Thomas S. Kuhn on a Draft of The Structure of Scientific Revolutions", in: *Studies in History and Philosophy of Science*, 26, 1995 [geschrieben 1960/61], S. 353-387; Paul Feyerabend, „Realism and Instrumentalism: Comments on the Logic of Factual Support", a.a.O., S. 245 und 251; Paul Feyerabend, „Problems of Microphysics", a.a.O., S. 145, Paul Feyerabend, „Consolations for the Specialist", in: Imre Lakatos und Alan Musgrave (Hrsg.), *Criticism and the Growth of Knowledge*. Cambridge: Cambridge University Press 1970, S. 197-230; und Paul Feyerabend, *Science in a Free Society*, a.a.O., S. 45. Zur Diskussion siehe auch Paul Hoyningen-Huene, „Feyerabends Kritik an Kuhns normaler Wissenschaft", in: J. Nida-Rümelin (Hrsg.), *Rationalität, Realismus, Revision*. Berlin: Walter de Gruyter 1999, S. 465-470; und Paul Hoyningen-Huene, „Paul Feyerabend und Thomas Kuhn", in: *Journal for General Philosophy of Science* 33 (1): 61-83 (2002).
29. Siehe z.B. Paul Feyerabend, „Consolations for the Spezialist", a.a.O.
30. Man mag zögern, hier den Begriff der Erkenntnistheorie zu verwenden, weil es sich um zwei relativ schlichte Forderungen handelt, die nicht den Grad von Elaboriertheit aufweisen, den man normalerweise mit dem Ausdruck Theorie verbindet.
31. Dies ist ein grosser Streitpunkt in der Diskussion mit Kuhn, der einen gewissen Dogmatismus in der normalen Wissenschaft positiv bewertet. Siehe Paul Feyerabend, „Two Letters of Paul Feyerabend to Thomas S. Kuhn on a Draft of The Structure of Scientific Revolutions", in: *Studies in History and Philosophy of Science*, a.a.O. Zur Diskussion, siehe Paul Hoyningen-Huene, „Feyerabends Kritik an Kuhns normaler Wissenschaft", a.a.O.; Paul Hoyningen-Huene, „Paul Feyerabend und Thomas Kuhn", a.a.O.; und Paul Hoyningen-Huene, „Three biographies: Kuhn, Feyerabend, and Incommensurability", in: Randy Harris (Hrsg.), *Rhetoric and Incommensurability*, New York: SUNY Press, im Druck.
32. Siehe z.B. Paul Feyerabend, „Explanation, Reduction and Empiricism", a.a.O.; und Paul Feyerabend, „Consolations for the Specialist", a.a.O.

33. Siehe z.B. Paul Feyerabend, „Professor Bohm's Philosophy of Nature. Review of *Causality and Chance in Modern Physics*. By David Bohm", in: *British Journal for the Philosophy of Science* 10, 1960, S. 321-338, hier S. 325 und 328; Paul Feyerabend, „Two Letters of Paul Feyerabend to Thomas S. Kuhn on a Draft of The Structure of Scientific Revolutions", a.a.O., S. 364-365; Paul Feyerabend, „Explanation, Reduction and Empiricism", a.a.O., S. 65; Paul Feyerabend, „How to Be a Good Empiricist: A Plea for Tolerance in Matters Epistemological", a.a.O.; Paul Feyerabend, „Review of *Scientific Change*. Edited by A. C. Crombie", in: *British Journal for the Philosophy of Science* 15, 1964, S. 244-254, hier S. 253-254; Paul Feyerabend, „A Note on the Problem of Induction", in: *The Journal of Philosophy* 61, 1964, S. 349-353, hier S. 351-353; Paul Feyerabend, „Problems of Empiricism", a.a.O., S.175 und 249-250, n. 121 und n. 122; Paul Feyerabend, „Reply to Criticism. Comments on Smart, Sellars and Putnam", a.a.O., S. 224 und 249, n. 25; Paul Feyerabend, „Review of *The Structure of Science*. By Ernest Nagel", in: *British Journal for the Philosophy of Science* 17, 1966, S. 237-249, hier S. 246-247; Paul Feyerabend, „Science, Freedom, and the Good Life", in: *The Philosophical Forum* 1, 1968, S. 127-135, hier 132; Paul Feyerabend, „Linguistic Arguments and Scientific Method", a.a.O., S. 58-59; Paul Feyerabend, „Consolations for the Spezialist", a.a.O., S. 208, Paul Feyerabend, *Against Method. Outline of an Anarchistic Theory of Knowledge*. London: New Left Books 1975, S. 38, 43, n. 9; und Paul Feyerabend, *Against Method. Outline of an Anarchistic Theory of Knowledge*. New York: Verso 1988. S. 27.
34. Feyerabends Standardbeispiel ist die Brownsche Bewegung, deren kritischer Charakter für die phänomenologische Thermodynamik erst im Lichte der statistischen Mechanik zu Tage tritt (siehe Endnote 35). Ähnliches gilt auch für den Rest von 43"/Jahrhundert von der gesamten Perihelpräzession des Merkur von 574"/Jahrhundert, die mittels der Newtonschen Gravitationstheorie nicht erklärbar zu sein schienen. Erst im Lichte der allgemeinen Relativitätstheorie wurden diese 43"/Jahrhundert zu einer widerlegenden Instanz der klassischen Gravitationstheorie.
35. Siehe dazu Paul Feyerabend, *Killing Time. The Autobiography of Paul Feyerabend*, a.a.O., S. 56, 65; Paul Feyerabend, *Science in a Free Society*, a.a.O., S. 109-111; und Paul Feyerabend und J. Ferber, „Single Magnetic Northpoles and Southpoles and Their Importance for Science – Ten Lectures delivered at the University of Vienna during the summer semester of 1947 by Dr. Felix Ehrenhaft, U.S. Visiting Professor. Compiled with the assistance of Professor Ehrenhaft and Dr. Schedling", Die Einführung und zwei Anhänge von Paul Feyerabend. Unveröffentlichtes Manuskript, 1947/1967, Anhang 2, S. 1-3. Feyerabend Archiv, Universität Konstanz: PF: 5-33.
36. Siehe z.B. Paul Feyerabend, „Dialogue on Method", in: G. Radnitzky und G. Andersson (Hrsg.), *The Structure and Development of Science*. Dordrecht: D. Reidel Pub. Co. 1979, S. 63-64; Paul Feyerabend, *Farewell to Reason*. London: Verso, 1987, S. 312; Paul Feyerabend, *Killing Time. The Autobiography of Paul Feyerabend*, a.a.O., S. 97, 101; und Paul Feyerabend, *Against Method. Outline of an Anarchistic Theory of Knowledge*. London: Verso 1993.
37. Vgl. Paul Feyerabend, „Reply to Criticism. Comments on Smart, Sellars and Putnam", a.a.O. S. 251, n. 1, mit dem Wiederabdruck Paul Feyerabend, *Realism, rationalism and scientific method. Philosophical papers*. Cambridge: Cambridge University Press 1981, S. 104, n. 1.
38. Dass Feyerabend, wie oben ausgeführt, diese Thesen auch selbst verwendet, ist kein Widerspruch, denn im Rahmen der immanenten Kritikstrategie lässt sie sich eventuell auch *gegen* kritische Rationalisten einsetzen.

Literatur

Baum, Wilhelm. (Hrsg.), *Paul Feyerabend – Hans Albert. Briefwechsel.* Frankfurt am Main: Fischer 1997.
Duhem, Pierre. *Ziel und Struktur der physikalischen Theorien.* Hamburg: Meiner 1978 [1906].
Feyerabend, Paul und J. Ferber, „Single Magnetic Northpoles and Southpoles and Their Importance for Science – Ten Lectures delivered at the University of Vienna during the summer semester of 1947 by Dr. Felix Ehrenhaft, U.S. Visiting Professor. Compiled with the assistance of Professor Ehrenhaft and Dr. Schedling", Die Einführung und zwei Anhänge von Paul Feyerabend. Unveröffentliche Manuskript, 1947/1967. Feyerabend Archiv, Universität Konstanz: PF: 5-33.
Feyerabend, Paul. *Zur Theorie der Basissätze.* Universitätsbibliothek Wien 1951.
Feyerabend, Paul. „Review of *Kant's Metaphysics and Theory of Science.* By Gottfried Martin", in: *British Journal for the Philosophy of Science* 7, 1956, S. 260-263.
Feyerabend, Paul. „An Attempt at a Realistic Interpretation of Experience", in: *Proceedings of the Aristotelian Society* 58, 1958, S. 143-170.
Feyerabend, Paul. „Complementarity", in: *Proceedings of the Aristotelian Society, 58, Suppl. Vol. 32*, 1958, 75-104.
Feyerabend, Paul: „On the Interpretation of Scientific Theories", in: *Proceedings of the 12th International Congress of Philosophy, Venice, 1958, Vol. 5.* Florence: Sansoni Editore 1960, S. 151-159.
Feyerabend, Paul. „Professor Bohm's Philosophy of Nature. Review of *Causality and Chance in Modern Physics.* By David Bohm", in: *British Journal for the Philosophy of Science* 10, 1960, S. 321-338.
Feyerabend, Paul. „Niels Bohr's Interpretation of the Quantum Theory", in: Herbert Feigl und Grover Maxwell (Hrgs.), *Current Issues in the Philosophy of Science. Symposia of Scientists and Philosophers. Proceedings of Section L of the American Association for the Advancement of Science, 1959.* New York: Holt, Rinehart and Winston 1961, S. 371-390.
Feyerabend, Paul. „Explanation, Reduction and Empiricism", In: Herbert Feigl und Grover Maxwell (Hrsg.), *Scientific Explanation,*

Space, and Time. Minnesota Studies in the Philosophy of Science, Vol. III. Minneapolis: University of Minnesota Press, 1962, S. 28-97.
Feyerabend, Paul. „Problems of Microphysics", in: R. Colodny (Hrsg.), *Frontiers of Science and Philosophy: University of Pittsburgh Series in the Philosophy of Science Vol. I*. Pittsburgh: University of Pittsburgh Press 1962, S. 189-283.
Feyerabend, Paul. „Review of *Erkenntnislehre*. By V. Kraft", in: *British Journal for the Philosophy of Science* 13, 1963, 319-323.
Feyerabend, Paul. „Materialism and the Mind-Body Problem", in: *The Review of Metaphysics* 17, 1963, S. 49-66.
Feyerabend, Paul. „Mental Events and the Brain", in: *The Journal of Philosophy* 60, 1963, S. 295-296.
Feyerabend, Paul. „How to Be a Good Empiricist: A Plea for Tolerance in Matters Epistemological", in: B. Baumrin (Hrsg.), *Philosophy of Science: The Delaware Seminar*. New York: Interscience Press 1963, S. 3-39.
Feyerabend, Paul. „Realism and Instrumentalism: Comments on the Logic of Factual Support", in: Mario Bunge (Hrsg.), *The Critical Approach to Science and Philosophy: In Honor of Karl R. Popper*. London: The Free Press of Glencoe 1964, S. 280-308.
Feyerabend, Paul. „Review of *Scientific Change*. Edited by A. C. Crombie", in: *British Journal for the Philosophy of Science* 15, 1964, S. 244-254.
Feyerabend, Paul. „A Note on the Problem of Induction", in: *The Journal of Philosophy* 61, 1964, S. 349-353.
Feyerabend, Paul. „Problems of Empiricism", in: R. Colodny (Hrsg.), *Beyond the Edge of Certainty. Essays in Contemporary Science and Philosophy*. Pittsburgh: CPS Publications in the Philosophy of Science 1965, S. 145-260.
Feyerabend, Paul. „Reply to Criticism. Comments on Smart, Sellars and Putnam", in: Robert Cohen und Marx Wartofsky (Hrsg.), *Proceedings of the Boston Colloquium for the Philosophy of Science 1962-64: In Honor of Philipp Frank. Boston Studies in the Philosophy of Science, Vol. II*. New York: Humanities Press 1965, S. 223-261.
Feyerabend, Paul. „Review of *The Structure of Science*. By Ernest Nagel", in: *British Journal for the Philosophy of Science* 17, 1966, S. 237-249.

Feyerabend, Paul. "Science, Freedom, and the Good Life", in: *The Philosophical Forum* 1, 1968, S. 127-135.
Feyerabend, Paul. "Linguistic Arguments and Scientific Method", in: *Telos* 2, 1969, S. 43-63.
Feyerabend, Paul. "Against Method: Outline of an Anarchistic Theory of Knowledge", in: M. Radner und S. Winokur (Hrsg.), *Analysis of Theories and Methods of Physics and Psychology. Minnesota Studies in the Philosophy of Science vol. IV*. Minneapolis: University of Minnesota Press 1970, S. 17-130.
Feyerabend, Paul. "Consolations for the Specialist", in: Imre Lakatos und Alan Musgrave (Hrsg.), *Criticism and the Growth of Knowledge*. Cambridge: Cambridge University Press 1970, S. 197-230.
Feyerabend, Paul. *Against Method. Outline of an Anarchistic Theory of Knowledge*. London: New Left Books 1975.
Feyerabend, Paul. *Science in a Free Society*. London: New Left Books 1978.
Feyerabend, Paul. "Dialogue on Method", in: G. Radnitzky und G. Andersson (Hrsg.), *The Structure and Development of Science*. Dordrecht: D. Reidel Pub. Co. 1979, S. 63-64.
Feyerabend, Paul. *Realism, rationalism and scientific method. Philosophical papers*. Cambridge: Cambridge University Press 1981.
Feyerabend, Paul. *Farewell to Reason*. London: Verso, 1987.
Feyerabend, Paul. *Against Method. Outline of an Anarchistic Theory of Knowledge*. New York: Verso 1988.
Feyerabend, Paul. *Against Method. Outline of an Anarchistic Theory of Knowledge*. London: Verso 1993.
Feyerabend, Paul. "Two Letters of Paul Feyerabend to Thomas S. Kuhn on a Draft of The Structure of Scientific Revolutions", in: *Studies in History and Philosophy of Science*, 26, 1995 [geschrieben 1960/61], S. 353-387.
Feyerabend, Paul. *Killing Time. The Autobiography of Paul Feyerabend*. Chicago: University of Chicago Press 1995.
Hoyningen-Huene, Paul. "Paul K. Feyerabend", in: *Journal for General Philosophy of Science* 28, 1, 1997, S. 1-18.
Hoyningen-Huene, Paul. "Feyerabends Kritik an Kuhns normaler Wissenschaft", in: J. Nida-Rümelin (Hrsg.), *Rationalität, Realismus, Revision*. Berlin: Walter de Gruyter 1999, S. 465-470.
Hoyningen-Huene, Paul. "Paul Feyerabend und Thomas Kuhn", in: *Journal for General Philosophy of Science* 33 (1): 61-83 (2002).

Hoyningen-Huene, Paul. „Three biographies: Kuhn, Feyerabend, and Incommensurability", in: Randy Harris (Hrsg.), *Rhetoric and Incommensurability*, West Lafayette: Parlor Press, im Druck.
Köhler, Wolfgang. „Physical Gestalten", in: W. Dennis (Hrsg.), *Readings in the History of Psychology*. New York: Appleton-Century-Crofts 1948.
Nagel, Ernst. *The Structure of Science. Problems in the Logic of Scientific Explanation*. London: Routledge 1966 [1961].
Oberheim, Eric. *On Feyerabend's Early Philosophy*. Dissertation. Universität Hannover, 2004.
Oberheim, Eric. „On the Historical Origins of the Contemporary Notion of Incommensurability: Paul Feyerabend's Assault on Conceptual Conservativism", in: *Studies in the History and Philosophy of Science*. Im Druck.
Oberheim, Eric und Paul Hoyningen-Huene. „Review Symposium: Radical Fallibilism vs. Conceptual Analysis: The Significance of Feyerabend's Philosophy of Science", in: *Metascience. The Australian Journal for The Philosophy of Science* 8, 2, 1998, S. 226-233.
Preston, John. *Feyerabend. Philosophy, Science and Society*. Oxford: Blackwell 1997.
Preston, John. „Feyerabend's Retreat from Realism", *Philosophy of Science* 64, 1997, S. 421-431.
Preston, John. „Review Symposia: Radical Fallibilism vs. Conceptual Analysis: The significance of Feyerabend's Philosophy of Science. Author's Response", in: *Metascience* 8, 2, 1998, S. 233-243.
Sankey, Howard. *The Incommensurability Thesis*. Aldershot: Avebury 1997.
Sankey, Howard. „Incommensurability: The Current State of Play", in: *Theoria* 12, 1997, S. 425-445.
Wittgenstein, Ludwig. *Philosophical Investigations*. Oxford: Basil Blackwell 1953.

ERHARD OESER

PAUL FEYERABEND ZWISCHEN WISSENSCHAFTSGESCHICHTE UND WISSENSCHAFTSTHEORIE

Vorbemerkung

Paul Feyerabend habe ich im Unterschied zu Viktor Kraft und Popper nie persönlich kennen gelernt. Auch habe ich nie einen Vortrag von ihm gehört. Als ich nach einem Studium in München bei Stegmüller 1960 nach Wien kam, um dort mein Studium zu beenden, war von Feyerabend nicht die Rede. Zum ersten Mal ist mir sein Name untergekommen, als ich, von einer längeren Lehrtätigkeit an der Universität Freiburg i. Br. nach Wien zurückgekehrt, von seinem spektakulären Vortrag „Die Wissenschaftstheorie, eine bisher unbekannte Form des Irrsinns" gehört hatte, den er auf dem deutschen Philosophiekongress gehalten hatte. Mit diesem provokativen Titel wollte er den Vertretern verschiedener Richtungen der Wissenschaftstheorie, die sich in ihrem Systematisierungsbestreben so eigenständig weiter entwickelt hatte, dass sie in vieler Hinsicht den Kontakt zur Naturwissenschaft selbst verloren hatte, einen heilsamen Schock versetzen. Als dann wenig später 1976 sein Buch *Against Method* in deutscher Sprache erschien, in dem er durch konkrete Fallstudien aus der Wissenschaftsgeschichte den Glauben an eine „allein selig machende" Verfahrensweise der wissenschaftlichen Erkenntnis erschüttern wollte, war das für mich der Anlass, mich ausführlich mit diesem Buch zu beschäftigen.

Das Ergebnis war eine dreiteilige Rundfunksendung, die zusammen mit einem Artikel von Stegmüller über die Kuhn-Popper-Debatte in der von Leo Gabriel herausgegebenen Zeitschrift *Wissenschaft und Weltbild* noch im selben Jahr erschien.[1] Ich selbst war – und bin es noch bis heute – ein Anhänger von Whewell, Mach und Boltzmann und betrachtete daher die Kritik Feyerabends an der formalistisch-normativen Wissenschaftstheorie als eine schon längst notwendige Rückkehr zur Wissenschaftsgeschichte, die ich sowohl als kritische Kontrollinstanz als auch als heuristisches „erkenntnistheoretisches Laboratorium" ansehe, um ein Wort von Dijksterhuis zu gebrauchen.

Dazu gehört aber auch für Feyerabend selbst nicht nur die Kenntnis der Geschichte der Naturwissenschaften sondern auch die der Philosophie und besonders der Erkenntnistheorie. Deshalb bricht auch Feyerabend für Aristoteles eine Lanze, an dem Popper in seiner Obsession auf die Vorsokratiker kein gutes Haar lässt. Feyerabend verteidigt Aristoteles, indem er die Erkenntnistheorie des Alltagsverstandes verteidigt und zeigt, dass Aristoteles kein Induktivist im Sinn der Popperianer ist und dass ihm daher auch die Schwierigkeiten des „Induktivismus" nicht zu schaffen machen.[2]

Auf die Frage „Was ist Erfahrung?" gibt die Aristotelische Philosophie nach Feyerabend eine einfache Antwort zur Hand: „Erfahrung – das ist, was uns unsere Sinne unter normalen Umständen (Tageslicht, wacher Beobachter; Abwesenheit ungewöhnlicher physikalischer Bedingungen) vermitteln, und was wir in der Alltagssprache beschreiben und an andere weitergeben. Die so definierte Erfahrung ist außerdem vertrauenswürdig, denn die Elemente des Universums sind (im Großen und Ganzen) harmonisch aufeinander abgestimmt. Harmonie herrscht insbesondere zwischen den Sinnesorganen eines Lebewesens und ihrer objektiven Funktion, die im Falle der Wahrnehmung in der korrekten Darstellung der Umwelt besteht. Wir können heute die allgemeinen Züge dieser Harmonie evolutionistisch erklären und damit weiter bestätigen".[3] Vor allem bei Boltzmann findet Feyerabend diese evolutionistische Erklärung: „Boltzmann wandte Darwinistische Prinzipien auf die Erkenntnis an und erklärte auch die Gesetze der Logik und Mathematik für nur vorläufig und der Verbesserung bedürftig. Hierin war er Popper weit voraus". Kritische Rationalisten sind für Feyerabend historische Analphabeten, und so entsteht bei ihnen die Idee der großen Originalität Poppers und der vielfachen Abhängigkeit von ihm.[4] Außerdem bedeutet für Feyerabend

> nicht jede Verwandtschaft mit Poppers Ideen Annahme der Popperschen Philosophie und Beeinflussung durch sie. Die Poppersche Philosophie ist aus vielen vorpopperschen Quellen zusammengerafft, gibt aber diese Quellen nur selten an. So scheint ein Schreiber, der aus denselben Quellen schöpft, für historische Analphabeten schon sofort ein Popperianer zu sein.

Die Rückkehr nicht nur zur Geschichte der Naturwissenschaft sondern auch zur Aristotelische Erkenntnistheorie und ihre evolutionistische Umdeutung im Sinne von Machs und Boltzmanns evolutionärer Erkenntnistheorie sind Züge in Feyerabends Wissenschaftstheorie, die doch ein anderes Bild von ihm erkennen lassen als das eines skurrilen erkenntnistheoretischen Anarchisten. In diesem Sinn sollen nun die nachfolgenden Bemerkungen verstanden werden, die sowohl Kritik als auch Rechtfertigung seiner Auffassung von Wissenschaftstheorie und Wissenschaftsgeschichte bedeuten.

Feyerabends Kritik an der zeitgenössischen Wissenschaftstheorie

Feyerabends Kritik ist meiner Ansicht nach primär nicht gegen die naturwissenschaftliche Methode als solche gerichtet, wie der Titel seines Buches *Against Method* nahe legen könnte, sondern vor allem gegen jenen „Methodenzwang", den der Wissenschaftstheoretiker durch normative Festlegungen auf den Naturwissenschaftler auszuüben versucht.

Ansätze zu dieser Kritik hatte Feyerabend schon vorher entwickelt. Darauf hatte bereits Viktor Kraft in der 2. Auflage seines Buches über den Wiener Kreis hingewiesen. Viktor Kraft hat jedoch diese Kritik Feyerabends an der Konsistenzforderung bei Theorien und der Bedeutungsinvarianz bei Begriffen lediglich als eine Modifikation des Empirismus angesehen.[5] Für die „anarchistische Erkenntnistheorie" und ihre polemischen Übertreibungen hatte er, als ich ihn darüber befragte, nur ein mitleidiges Lächeln übrig.

Mit der Berufung auf die faktische Wissenschaftsgeschichte glaubte jedoch Feyerabend eine Waffe in der Hand zu haben, mit der es möglich ist, die normativen Ansprüche der geläufigen Wissenschaftstheorie überhaupt zu vernichten. Denn in der faktischen Wissenschaftsgeschichte zeigt sich für ihn, dass sich wahre wissenschaftliche Erfolge nur dann eingestellt haben, wenn gerade die geläufigen Regeln wissenschaftlichen Handelns verletzt worden sind, die den Fortschritt nur behindert hätten. Diese Fallstudien bilden nach Feyerabends eigener Ansicht das Zentrum seines Buches:

> Sie sind das Material, an dem die Mängel eines abstrakt-rationalen Vorgehens erläutert und terminologisch fixiert werden. Die mehr abstrakten Erörterungen und die Polemik gegen den Rationalismus sind ohne dieses Material und ohne diese Erläuterungen nicht denkbar. Sie sind durchaus sekundär.[6]

An diese Charakterisierung, die Feyerabend allerdings erst nachträglich in der deutschen Ausgabe seines Werkes eingefügt hat, muss man sich halten, wenn man die bewusst polemisch und provokant formulierten Thesen seines erkenntnistheoretischen Anarchismus, den er ausdrücklich nicht mit dem politischen Anarchismus gleichsetzt, überhaupt ernst nehmen will.

Feyerabend beginnt seine Untersuchung mit einem Motto von Bertolt Brecht: „Ordnung ist heutzutage meistens dort, wo nichts ist".[7] Übertragen auf das Phänomen „Wissenschaft" bedeutet diese Aussage für Feyerabend, dass die „Gesetze der Vernunft", auf die sich die erkenntnistheoretisch-methodologischen Vorschriften stützen, geradezu eine „verdummende Wirkung" hervorrufen. Sie simplifizieren die Wissenschaft, indem sie diejenigen simplifizieren, die sie betreiben. Der Vorgang ist stets der gleiche: Zunächst wird ein bestimmtes Forschungsgebiet festgelegt, von der Entwicklung der anderen Gebiete abgetrennt und mit einer eigenen „Logik" ausgestattet. Eine gründliche Ausbildung in einer solchen Logik bestimmt dann das Arbeiten auf dem fachlich unbegrenzten Gebiet. Sie vereinheitlicht die Handlungen und bringt auch große Teile des Geschichtsablaufs zum Stillstand.

Feyerabend leugnet mit dieser Charakterisierung der abstrakt normativen Wissenschaftstheorie nicht, dass man damit eine Tradition schaffen kann, die durch strenge Regeln zusammengehalten wird und die auch gewisse Erfolge aufzuweisen hat. Seine Frage ist vielmehr, ob es wünschenswert ist, eine solche Tradition zu unterstützen und alles andere auszuschließen. Seine Antwort darauf lautet: NEIN. Und das zeigt die reale Wissenschaftsgeschichte. Denn sie ist nach Feyerabend voll von Beispielen einer ständigen Verwerfung starrer Regeln. Sie besteht nicht bloß aus Tatsachen und Schlüssen aus Tatsachen, sondern ist vielmehr so komplex, chaotisch voll von Fehlern und so unterhaltend wie die in ihr enthaltenen Ideen. Verzichtet man aber auf eine Simplifizierung der Wissenschaftsgeschichte, dann gibt es für Feyerabend nur eine adäquate Charakterisierung des realen Phänomens Wissenschaft: „Die Wissenschaft ist wesentlich ein anarchistisches Unternehmen: Der theoretische Anarchismus ist menschenfreundlicher und eher geeignet, zum Fortschritt anzuregen, als Gesetz- und Ordnungs-Konzeptionen"[8].

Es besteht aber nach Feyerabend kein Grund zu der Befürchtung, dass ein solcher Anarchismus zum Chaos führt. Dazu ist das menschliche Nervensystem zu hoch organisiert. Damit wird jedoch auch zugegeben, dass der Wissenschaft die Suche nach Regeln und Methoden zur Sicherung der Erkenntnis notwendig immanent ist. Das bedeutet aber, dass der erkenntnistheoretische Anarchismus selbst einer Methode bedarf, um überhaupt Zutritt zur Wissenschaft zu bekommen. Diese Methode ist die Methode der Antiregeln.

Die Methode der Antiregeln

Diese Methode besteht im wesentlichen darin, folgendes zu zeigen: „In jeder Regel, sei sie noch so grundlegend oder notwendig für die Wissenschaft, gibt es Umstände, unter denen es angezeigt ist, die Regel nicht nur zu missachten, sondern ihrem Gegenteil zu folgen." Der Anarchismus reduziert sich damit zu einer relativ harmlosen Befürwortung einer liberalen Anwendungspraxis gültiger methodologischer Regeln. Vor allem dann, wenn diese Praxis der Antiregeln auf eine bestimmte Phase beschränkt bleibt. Diese Phase wird von Feyerabend als ein unvernünftiges, unsinniges, unmethodisches Vorspiel angesehen, das sich als unerlässliche Vorbedingung der Klarheit und des empirischen Erfolgs erweist. Jede noch so strenge, exakte, logisch konsistente Wissenschaft wird nur dann Erfolge haben, wenn gelegentlich anarchistische Schritte zugelassen werden.

Diese Auffassung unterscheidet sich allerdings nur wenig von den bekannten Überlegungen Kuhns zur Struktur wissenschaftlicher Revolution. Radikalisiert wird sie erst durch die Festlegung eines Grundsatzes, der die wissenschaftliche Revolution zu einem Permanenzzustand macht. Dieser Grundsatz, der sich nach Feyerabend unter allen Umständen und in allen Stadien der menschlichen Entwicklung vertreten lässt, lautet: „Anything goes (mach, was du willst)". Es ist, wie Feyerabend selbst ausdrücklich feststellt, ein „abstrakter Grundsatz", der in „seinen konkreten Einzelheiten untersucht und erklärt" werden muss.[9] Er bedeutet nicht, dass es überhaupt keine allgemeinen Grundsätze in der Wissenschaft gibt, sondern dass sich derartige Grundsätze von Fall zu Fall verändern können und müssen. Praktisch heißt das, dass ein Wissenschaftler darauf angewiesen ist, sein Problem gemäß der vorliegenden Situation zu behandeln. Er muss die besonderen Umstände berücksichtigen und darf seine Aufgabe nicht als gelöst betrachten, nur weil eine Lösung vorliegt, die einem allgemeinen Grundsatz entspricht. Der einzige allgemeine Grundsatz, der immer gilt, bildet selbst keine Lösung und Hilfe in diesem ständig erneuerten Prozess der Problemstellungen. Er dient vielmehr dazu, das Ganze in Bewegung zu halten. Den Grundsatz „Mach, was du willst" in seinen konkreten Einzelheiten zu untersuchen, heißt für Feyerabend, die „Konsequenzen von ‚Antiregeln' zu verfolgen, die gewissen bekannten Regeln des wissenschaftlichen Vorgehens entgegengesetzt sind".[10] Feyerabend tritt mit dieser Zielsetzung gegen die herrschende wissenschaftstheoretische Auffassung auf, dass der Erfolg unserer Theorie durch „Erfahrung" oder die „Tatsachen" oder durch „experimentelle Ergebnisse" gemessen wird. Übereinstimmung zwischen Theorie und „Daten" spricht für die Theorie, während Nichtübereinstimmung sie erschüttert und womöglich zu ihrer Aufgabe zwingt. Die entsprechenden „Antiregeln", die Feyerabend dagegen aufstellt, weisen uns an, Hypothesen einzuführen und auszubauen, die gut bestätigten Theorien oder Tatsachen widersprechen.[11]

Die erste spezielle Antiregel ist gegen die so genannte „Konsistenzbedingung" gerichtet. Diese Bedingung, nach der neue Hypothesen mit bereits anerkannten Theorien übereinstimmen sollen, ist für Feyerabend deswegen unvernünftig, weil sie nur die ältere und nicht die bessere Theorie am Leben hält.[12] Nur eine Vielfalt von einander widersprechenden Alternativtheorien erweist sich für die Entwicklung der Wissenschaft als fruchtbar. Alternativen nur dann zuzulassen, wenn die alte Theorie bereits durch empirische Daten erschüttert ist, ist vor allem deswegen nicht praktikabel, weil gerade diese Daten oft nur mit Hilfe einer der alten Theorie widersprechenden Alternative zutage gefördert werden können. Auch erkennt man einige der wichtigsten formalen Eigenschaften einer Theorie nicht durch Analyse, sondern durch Kontrast. Das heißt, der Wissenschaftler muss eine pluralistische Methodologie verwenden, er muss Ideen mit Ideen vergleichen und nicht nur mit der Erfahrung. Wenn er dies tut, wird er nach Meinung Feyerabends begreifen müssen, dass die wissenschaftliche Erkenntnis in diesem Sinne keine Abfolge in sich widerspruchsfreier Theorien ist, die gegen eine Idealtheorie hin konvergieren, sie ist keine allmähliche Annäherung an die Wahrheit. Sie ist vielmehr ein stets anwachsendes Meer miteinander unverträglicher Alternativen; jede einzelne Theorie, jedes Märchen, jeder Mythos, der dazu gehört, zwingt die anderen zu deutlicherer Entfaltung, und alle tragen durch ihre Konkurrenz zur Entwicklung unseres Bewusstseins bei. Nie ist etwas endgültig ausgemacht, keine Auffassung kann je aus einer umfassenden Darstellung weggelassen werden. Was Feyerabend damit zu recht kritisiert, ist die Approximationstheorie der Wahrheit von Popper, was er aber dabei übersieht, ist die Konsequenz einer Darwinistischen Erkenntnistheorie, die er gerade bei Boltzmann befürwortet hat. Denn das Meer der miteinander unverträglichen Alternativen wird auf drastische Weise durch das Selektionsprinzip eingeschränkt. In diesem Sinne steht die Falsifikationstheorie Poppers der evolutionären Erkenntnistheorie Machs und Boltzmanns näher als Feyerabends uneingeschränkter Proliferation von Alternativtheorien.

Die zweite Antiregel Feyerabends, die nicht nur zugunsten von Hypothesen spricht, die zu bereits bestehenden Theorien in Widerspruch treten, sondern auch Hypothesen befürwortet, die in direkter Weise Beobachtungen, Tatsachen und experimentellen Ergebnissen widersprechen, scheint radikaler zu sein, als sie es in Wirklichkeit ist. Denn bisher hat es noch keine Theorie gegeben, die mit allen bekannten Tatsachen auf ihrem Gebiet völlig übereingestimmt hätte. Die Frage wird daher von Feyerabend mit Recht auf das Problem eingeschränkt, ob die unweigerlich bestehenden Unstimmigkeiten zwischen Theorie und Tatsachen vermehrt oder vermindert werden sollen oder was sonst mit ihnen geschehen soll. Die Antwort darauf wird von Feyerabend konsequent im Sinne seines erkenntnistheoretischen Anarchismus gegeben, der Antiregeln aufstellt, um die Wissenschaft in Bewegung zu halten. Denn das heißt, den „Fortschritt" ständig neu zu involvieren. Die zweite Antiregel versucht daher die Unstimmigkeit zwischen Theorie und Erfahrung nicht abzuschwächen, sondern zu vermehren. Sie lautet: „Man muss ein neues Begriffssystem erfinden, das mit den bestfundierten Beobachtungsergebnissen in Konflikt steht, sie außer Kraft setzt, das die einleuchtendsten theoretischen Grundsätze durcheinander bringt und Wahrnehmungen einführt, die nicht in die bestehende Wahrnehmungswelt passen." Gerechtfertigt wird diese Antiregel von Feyerabend damit, dass es nach seiner Auffassung so etwas wie theorieunabhängige Tatsachen nicht gibt. Die Verknüpfung von Theorien und Tatsachen ist viel enger als man gemeinhin annimmt. Nicht nur hängt die Beschreibung jeder einzelnen Tatsache von irgendeiner Theorie ab, sondern es gibt Tatsachen, die überhaupt nicht ins Blickfeld des Wissenschaftlers geraten, wenn er nicht eine zur gängigen bestehenden Theorie in Widerspruch stehende Alternative ersinnt. Das deutet daraufhin, dass die methodologische Einheit, auf die man sich bei der Diskussion von Fragen der Prüfung und des empirischen Gehaltes beziehen muss, aus einer ganzen Menge sich teilweise überschneidender, mit den Tatsachen vereinbarer aber miteinander unverträglicher Theorien besteht.

Diese Feststellungen klingen nüchtern und wenig revolutionär. Sie bedeuten für den Wissenschaftstheoretiker nicht mehr und nicht weniger, als dass er die klassischen vielleicht allzu einfachen Prüfungsmodelle durch ein komplexes Modell zu ersetzen hat. Um diesen Eindruck einer Ersetzung einer Methodologie durch eine andere, die lediglich komplexer und damit liberaler ist, zu entgehen, setzt Feyerabend an den Schluss einer allgemeinen Überlegung eine anarchistische Predigt:

Eine einheitliche Meinung mag das Richtige sein für die Kirche, für die eingeschüchterten oder gierigen Opfer eines (alten oder neuen) Mythos oder für die schwachen und willfährigen Untertanen eines Tyrannen. Für die objektive Erkenntnis brauchen wir viele verschiedene Ideen. Und eine Methode die die Vielfalt fördert ist auch als einzige mit einer humanistischen Auffassung vereinbar.[13]

Um die Vielfalt der Ideen und den darauf beruhenden Theorienpluralismus zu garantieren, muss Feyerabend eine zusätzliche allgemeine Regel aufstellen, die jedoch schon längst von Kant mit seiner These von der Unmöglichkeit des absoluten Irrtums vorweggenommen worden ist. Bei Kant findet sich diese Idee bereits in der „Allgemeinen Naturgeschichte und Theorie des Himmels" vom Jahre 1755. Sie lautet: „Auch in den aller unsinnigsten Meinungen, welche sich bei den Menschen haben Beifall erwerben können, wird man jederzeit etwas Wahres finden können".[14] Diese Regel ist für Feyerabend nur eine Konsequenz aus seinem allgemeinen Grundsatz „Mach, was du willst" und lautet bei ihm folgendermaßen: „Kein Gedanke ist so alt oder absurd, dass er nicht unser Wissen verbessern könnte".[15] Das heißt, dass man berechtigt ist, Alternativen, die zur Aufrechterhaltung des Wettstreits in der Wissenschaft dienen, auch aus der scheinbar durch den Fortschritt erledigten Vergangenheit zu entnehmen. Denn im Grunde genommen ist nach Feyerabend auch die fortgeschrittenste und scheinbar gesichertste Theorie nicht sicher davor, dass sie mit Hilfe von Auffassungen verändert oder gänzlich gestürzt werden kann, die eine hochmütige Unwissenheit schon in den Mülleimer der Geschichte geworfen hat. Alternativen soll man daher aufgreifen, wo immer man sie findet:

> Sei dies nun in der antiken Mythologie, in modernen Vorurteilen, in den Elaboraten von Fachleuten oder den Fantasien komischer Käuze. Die gesamte Geschichte einer Disziplin wird herangezogen, um ihren neuesten und ‚fortgeschrittensten' Entwicklungszustand zu verbessern.

Damit wird jedoch die Idee eines Abgrenzungskriteriums in jeder Hinsicht gänzlich aufgegeben. Der Liberalismus des Theorien- und Methodenpluralismus wird zur Anarchie. In Feyerabends eigenen Worten: „Die Trennung zwischen der Geschichte einer Wissenschaft, ihrer Philosophie und der Wissenschaft selber löst sich in nichts auf, desgleichen der Unterschied zwischen Wissenschaft und Nichtwissenschaft".[16]

Die Methode der Fallstudien: Galilei

Der erkenntnistheoretische Anarchismus, ausgedrückt durch den trivialen und inhaltsleeren Grundsatz: „Mach, was du willst", hätte nur wenig Chance, in der Diskussion um die Wissenschaftlichkeit der Wissenschaft ernst genommen zu werden, wenn Feyerabend nicht zur Stützung seiner Auffassung eine Reihe ausführlicher historischer Fallstudien geliefert hätte. Diese Fallstudien bilden nach seiner eigenen Auffassung das Zentrum seiner Arbeit und entsprechen seiner grundlegenden Leitidee, dass die Geschichte der Wissenschaft ein untrennbarer Teil der Wissenschaft selbst ist – übrigens eine Auffassung, die heutzutage von keinem Wissenschaftler und keinem Wissenschaftstheoretiker mehr geleugnet wird. Unter all den Naturwissenschaftlern der Neuzeit, deren Entdeckungen und Theorien Feyerabend zur Stützung seiner wissenschaftstheoretischen Überlegungen heranzieht, ragt ein Name immer wieder hervor: Galileo Galilei. Er ist für Feyerabend Repräsentant und Gewährsmann für eine Vorgangsweise, die dem erkenntnistheoretischen Anarchismus eine faktisch-historische Grundlage zu liefern vermag.

Gerade weil Galilei kein „methodologischer Normalverbraucher" war, erzielte er Fortschritte, indem er neue Begriffe einführte und gewohnte Verbindungen zwischen Wörtern veränderte. Beispiele für eine neue Interpretation des Begriffes der Bewegung ist einerseits die Bewegung eines Schiffssegels einmal vom Ufer aus beobachtet und einmal vom Schiff selbst aus beobachtet und andererseits das Turmargument, das die Gegner des kopernikanischen Systems verwendeten. Nach Copernicus müsste die Bewegung eines von einem Turm herabfallenden Steines „gemischt", „geradlinig" und „kreisförmig" sein. Die vom Beobachter wahrgenommene Bewegung ist aber lediglich geradlinig senkrecht nach unten. Dieses Ergebnis widerlegt aber die Kopernikanische Theorie nur dann, wenn der Begriff der Bewegung in der Beobachtungsaussage der gleiche ist, wie der in der Kopernikanischen Voraussage. Gerade das ist aber nicht der Fall. Im Kopernikanischen Begriffssystem ist mit dem Begriff „Bewegung des Steines" nicht lediglich eine Bewegung bezüglich eines Punktes im Gesichtsfeld eines Beobachters gemeint, d. h. die wahrgenommene Bewegung, sondern vielmehr die Bewegung des Steines im Sonnensystem oder im absoluten Raum überhaupt. Diese vom Beobachter auf der Erde als unabhängig zu betrachtende Bewegung ist dann die wirkliche Bewegung, deren Verlauf mit der Kopernikanischen Voraussage übereinstimmt. Dieser neue Bewegungsbegriff, der auf die Bewegung des Steines im

Sonnensystem und nicht auf den terrestrischen Beobachter bezogen ist, ist ein Begriff einer neuen höchst abstrakten Beobachtungssprache, die mit der natürlichen Alltagssprache nicht übereinstimmt. Der „psychologische Trick" besteht nun nach Feyerabend darin, dass Galilei diese Tatsache verschleiert, indem er dieses plausible Beispiel aus dem Alltagsleben auf die komplizierte Kopernikanische Theorie des Sonnensystems überträgt. Dann ergibt sich eine einfache Erklärung für die Unbeobachtbarkeit der wirklichen Bewegung des Steines in Bezug auf das Sonnensystem: Der Stein auf der Spitze des Turms bleibt für einen Beobachter am Fuße des Turms unbewegt, weil der Beobachter auf der Erde die Umdrehungsbewegung der Erde mitmacht. Genauso wenig wie der Passagier an Bord eines Schiffes die Bewegung des Segels erkennen kann, da er mit dem Segel und dem ganzen Schiff mitgeführt wird, genauso wenig kann der Beobachter auf der Erde die Bewegungen der Gegenstände auf der Erdoberfläche beobachten, die sich durch die Erdumdrehung ergeben.

Diese Vorgangsweise Galileis ist für Feyerabend keine Argumentation, sondern „Überredungskunst". Unter ihrem Einfluss beginnt man ganz automatisch, die Bedingungen der beiden Fälle durcheinander zu bringen, und wird Relativist. Galilei behält also im Kampf um das Kopernikanische System nach Feyerabend wegen seines Stils und seiner Überredungskunst die Oberhand; zusätzlich deswegen, weil er auf Italienisch und nicht nur auf Lateinisch schreibt; und weil er sich an Leute wendet, die gefühlsmäßig gegen die alten Ideen und die mit ihnen verbundenen Maßstäbe der Gelehrsamkeit eingenommen sind. Seine Fernrohrbeobachtungen, mit denen er die Kopernikanische Theorie stützen wollte, waren undeutlich und unbestimmt. Seine Zeichnungen der Mondoberfläche waren falsch. Man braucht heutzutage nur einen kurzen Blick auf Galileis Zeichnungen zu werfen und sie mit heutigen Fotografien entsprechender Phasen zu vergleichen, um sich zu überzeugen, dass „kein Bestandteil der Zeichnungen mit Sicherheit mit irgendeiner bekannten Stelle der Mondlandschaft identifiziert werden kann".[17] Die Entdeckung der Venusphasen zeichnete Galilei ebenfalls falsch. Außerdem bewiesen sie die Erdbewegung nicht und damit auch nicht das kopernikanische System. Aus dem Fall Galileis schließt Feyerabend, dass die Hinwendung zu neuen Ideen grundsätzlich nicht mit Argumenten bewirkt werden kann, sondern nur mit irrationalen Mitteln wie Propaganda, Gefühl, ad-hoc-Hypothesen und Berufung auf Vorurteile aller Art.

Was jedoch Feyerabend übersehen hat, ist die historische Tatsache, dass die Verteidigung und Weiterentwicklung des kopernikanischen Systems nicht Galileis große wissenschaftliche Leistung war. Sie ging eigentlich völlig daneben und bildete sogar ein Hindernis für die Akzeptanz der neuen Theorie des Sonnensystems, deren Wahrheit Galilei mit seiner falschen Gezeitentheorie zu beweisen versuchte. Denn nach seiner Erklärung war es allein die doppelte Erdbewegung der Erde, die das Ebbe-und-Flut-Phänomen hervorrufen soll, was jedoch zu einem zwölfstündigen Rhythmus führt und damit dem real beobachtbaren 6-Stundenrhythmus widerspricht.[18] Die eigentliche Leistung Galileis besteht in der Entdeckung der Fall- und Wurfgesetze, mit der er zum Begründer der neuzeitlichen Physik wurde. Die Begründung der nuova scienza erfolgte jedoch nach der strikten heuristischen Methode des metodo risolutivo und metodo compositivo, die jeweils das Resultat eines Experimentes zur Begründung des nächsten komplexeren Experimentes verwendet. Auf diese Weise ist Galilei über die Experimente zur Waage, zum Hebel und zur schiefen Ebene zum Gesetz des freien Falls und der Wurfbewegung gekommen.[19] Newton hat diese Methode der induktiven Analyse und konstruktiven Synthese zur Begründung der universalen Mechanik benützt und hat erst mit der Integration von Keplers Gesetzen, die Galilei selbst völlig ignoriert hat, und den terrestrischen Gesetzen der Fall- und Wurfbewegung dem kopernikanischen System zum Durchbruch verholfen. Dabei spielte vor allem auch seine Erklärung des Ebbe-und-Flut-Phänomens durch die Gravitationskräfte von Sonne und Mond eine entscheidende Rolle.

Die Konsequenz aus dieser Betrachtung des Falles Galilei ist, dass zur Überprüfung von konkurrierenden wissenschaftstheoretischen Auffassungen nicht nur isolierte historische Fallstudien ausreichen, sondern dass immer die Gesamtentwicklung eines Wissensgebietes von seinen Anfängen bis zur jeweiligen Gegenwart rekonstruiert werden muss, wie es z.B. Mach mit seiner Entwicklungsgeschichte der Mechanik oder wie es Whewell mit dem Gesamtgebiet der Naturwissenschaften getan hat. Nur auf diese Weise, d.h. „begründet auf der Wissenschaftsgeschichte",[20] lässt sich ein brauchbares Modell des wissenschaftlichen Erkenntnisfortschritts aufstellen, das sowohl den Dogmatismus der rein normativen Wissenschaftslogik als auch den erkenntnistheoretischen Anarchismus vermeidet.

So gesehen erweist sich auch die Darstellung des Falles Galilei durch Feyerabend als unzureichend und trügerisch. Denn Feyerabend berücksichtigt nur eine Seite in Galileis wissenschaftlichem Schaffen, die zwar die spektakulärste und bekannteste ist, aber keineswegs den wissenschaftlichen Ruhm Galileis begründete. Galileis wahre Leistung liegt sowohl nach dem Urteil seiner Zeitgenossen als auch seiner Nachfolger ganz eindeutig nicht auf dem Gebiet der Astronomie, sondern der terrestrischen Mechanik. Seine Entdeckung der Gesetze der Fall- und Wurfbewegungen führte zu einer neuen Grundlagentheorie, auf der die gesamte neuzeitliche Physik aufbauen konnte. Und gerade dieser Theorie hat Galilei selbst jene Form gegeben, die wohl als die strengste Form des Methodenzwanges angesehen werden kann: die Form eines axiomatisch-deduktiven Systems, in dem notwendig und zwangsweise ein Argument aus dem anderen folgt, ohne dass subjektive Willkür in diese geschlossene Kette eingreifen könnte. Wenn Galilei also kein erkenntnistheoretischer Anarchist im Sinne Feyerabends war, so war er auch andererseits kein „methodologischer Normalverbraucher". Denn er wusste genau, dass die Akzeptierung methodologischer Vorschriften und Regeln noch keinen absoluten Zwang bedeutet. Jedes System einer Wissenschaft nämlich, mag es noch so streng formalisiert und geschlossen sein, weist offene Stellen auf, an denen die neuen Ideen eindringen können, die schließlich das ganze System von Grund auf verändern. Auf diese Weise ist Galilei auf dem Gebiet der terrestrischen Bewegungslehre gelungen, was ihm in der Himmelsmechanik versagt geblieben ist. Die aristotelisch-scholastische Bewegungslehre konnte Galilei durch seine streng methodisch aufgebaute Theorie der Fall- und Wurfbewegungen überwinden. Die Ptolemäische Theorie des Sonnensystems jedoch hatte so lange ihren berechtigten Sinn, als es nicht eine ihr formal und methodisch gleichwertige Theorie gab, die sie an empirischem Gehalt, d. h. Erklärungswert und Voraussagefähigkeit, übertraf. Der Kampf um das Kopernikanische Weltsystem wurde daher auch nicht durch Galilei, sondern erst durch Kepler und Newton entschieden. Das Beispiel Galilei zeigt daher, dass die Wissenschaftsgeschichte zwar gerade wegen der komplexen Struktur der in ihr enthaltenen Ideen chaotisch und voller Fehler sein kann, dass aber in ihr trotzdem eine innere Gesetzmäßigkeit und methodische Eigendynamik wirksam ist, durch die immer wieder neue Ordnungen geschaffen und alte Fehler korrigiert werden.

Angesichts der Tatsache, dass die Wissenschaftstheorie des 19. Jahrhunderts bei Whewell, Duhem, Mach, Boltzmann u.a. nichts anders war als eine Rekonstruktion der Wissenschaftsgeschichte mit dem Ziel methodologische Regeln für den Fortschritt wissenschaftlicher Erkenntnis aufzudecken, und diese Rückkehr zur Wissenschaftsgeschichte bereits durch Kuhn und Lakatos u.a. ebenfalls propagiert worden ist, fragt sich vielmehr wie das Verhältnis von Wissenschaftstheorie und Wissenschaftsgeschichte zu verstehen ist und welchen Beitrag Feyerabend dazu geleistet hat.

Während diese Fragestellung für die genannten Autoren aus dem 19. und beginnenden 20. Jahrhundert einfach zu beantworten ist, weil sie beide Disziplinen nicht voneinander getrennt sahen, sondern in sich selbst vereinten und außerdem Fachleute jener Disziplinen waren, führte die moderne Rückkehr der Wissenschaftstheoretiker zur Wissenschaftsgeschichte zu polemischen Kontroversen sowohl innerhalb der Wissenschaftstheorie als auch zwischen Wissenschaftstheoretikern und Historikern. So sagt Paolo Rossi in einer Kritik an Feyerabend: „Die Historiker dürfen sich nicht zu Sammlern von Beispielfällen machen lassen, die die Wissenschaftstheoretiker dann als Belege für ihre theoretischen Konstruktionen benutzen".[21] Genau das aber macht Feyerabend, wenn er sich in seiner Fallstudie weniger selbst auf die Originalschriften als auf die Sekundärliteratur einer Reihe von Astronomiehistorikern wie Zinner, Wolf, Kopal, Schulz u.a. stützt, um zu zeigen, wie fehlerhaft Galileis Fernrohrbeobachtungen waren, und dass es gerade diese „Nachlässigkeit" war, die bewirkte, dass sich die moderne Wissenschaft rasch in die „richtige" Richtung entwickelt hat. Die Antwort Feyerabends auf Rossis Kritik, dass auch die Historiker eine Philosophie, allerdings eine primitive und falsche, unbewusst voraussetzen, geht jedoch an dem Kern der Fragestellung vorbei, der darin besteht, dass man in der Wissenschaftsgeschichte immer Beispiele für jede philosophische Position finden kann. Die einzig mögliche und von den Autoren des 19. Jahrhunderts auch schon praktizierte Lösung besteht darin, ein Wissensgebiet in der Gesamtheit seines historischen Ablaufes im Hinblick auf die dort verwendeten Methoden zu rekonstruieren, wodurch vorweggenommene normative methodologische Überlegungen auf ihre Brauchbarkeit überprüft werden und sich neue Einsichten in neue Methoden ergeben können.[22]

Gerechtfertigt ist Feyerabends „erkenntnistheoretischer Anarchismus" nur deswegen, weil Feyerabend in der historischen Situation der Entwicklung der Wissenschaftstheorie, die sich immer mehr von der tatsächlichen wissenschaftlichen Praxis entfernt hatte und zu einer „Buchprüfer"disziplin degeneriert war, eine Rückkehr zur Wissenschaftsgeschichte gefordert hatte, die in dieser polemischen Radikalität nicht zu übersehen war.

Anmerkungen

1. Erhard Oeser, „Der erkenntnistheoretische Anarchismus", in: *Wissenschaft und Weltbild*, 29. Jg. Nr. 3/4. 1976, S. 181-194.
2. Paul Feyerabend, „Eine Lanze für Aristoteles", in: G. Radnitzky und G. Andersson (Hrsg.), *Fortschritt und Rationalität der Wissenschaft.* Tübingen: Mohr 1980. S. 183.
3. Paul Feyerabend, Der *wissenschaftstheoretische Realismus und die Autorität der Wissenschaften.* Braunschweig: Vieweg 1978. S. 249.
4. Ebd., S. 28.
5. Viktor Kraft, *Der Wiener Kreis*. 2. Aufl., Wien New York: Springer 1968. S. 195.
6. Paul Feyerabend, *Wider den Methodenzwang. Skizze einer anarchistischen Erkenntnistheorie.* Frankfurt/Main: Suhrkamp 1976. S. 26.
7. Ebd., S. 28.
8. Ebd., S. 28.
9. Ebd., S. 46.
10. Ebd., S. 47.
11. Ebd., S. 47.
12. Ebd., S. 53.
13. Ebd., S. 67f.
14. Immanuel Kant, *Allgemeine Naturgeschichte und Theorie des Himmels,* 1755. In: Kants Werke. Akademie Textausgabe. Berlin 1968, Bd. I, S. 227.
15. Paul Feyerabend, *Wider den Methodenzwang,* a.a.O., S. 69.
16. Ebd., S. 70.
17. Ebd., S. 170.
18. Erhard Oeser, *Wissenschaftstheorie als Rekonstruktion der Wissenschaftsgeschichte.* 2. Bd. Wien, München: Oldenburg 1979. S. 96 ff.
19. Ebd. S. 7ff.
20. William Whewell, *The Philosophy of Inductive Science founded upon their History.* Reprint. London: Routledge/Thoemmes Press 1996.
21. Paolo Rossi, "Hermeticism and Rationality in the Scientific Revolution", in: M.L. Righini Bonelli und W. Shea (Hrsg.): *Reason, Experiment and Mysticism in the Scientific Revolution.* New York 1975, S. 248. Vgl. Paul Feyerabend, *Wider den Methodenzwang,* a.a.O., S. 195.
22. Erhard Oeser, *Popper, der Wiener Kreis und die Folgen. Die Grundlagendebatte der Wissenschaftstheorie.* Wien: Wiener Universitätsverlag 2003. S. 167f.

REINHOLD KNOLL

WARUM WISSENSCHAFT EINE KUNST IST ...
GEDANKEN ZU PAUL FEYERABEND

So man am Leben der Universität Wien nach dem Zweiten Weltkrieg schon Interesse zeigte, war es unvermeidbar, von allen möglichen Geschichten zu hören. Natürlich dominierten die Gespräche Informationen über strenge Prüfer der Studenten, diverse Eigenheiten der Lehrer, die karge Ausstattung der Bibliotheken, die mühevoll-merkwürdigen Verfahren über den Weiterverbleib von Lehrpersonal nach 1945 und vieles anderes mehr.[1] Allerdings blieb nachhaltig in Erinnerung, dass es im philosophischen Seminar hin und wieder zu heftigen Kontroversen gekommen sein soll, so nur ein Mann darin auftauchte, der regelmäßig den Seminarleiter zu irritieren vermochte und zum Ärger des Institutes die Teilnehmer offensichtlich so beeindruckte, dass sich in den Pausen und in den Kolloquia nach dem Seminar eine Menschentraube um ihn bildete, um mehr von seinen Ideen und Gedanken zu hören. Der Name hatte sich herumgesprochen: Paul Feyerabend.[2] Im Rückblick gewinnen diese Berichte eine Form der Bestätigung dafür, was der Leser seiner Schriften recht gut erkennen kann. In der weit wichtigeren und späteren Kontroverse zwischen Karl Popper und Thomas Kuhn schien es völlig überraschend auch eine dritte Position zu geben, eben Paul Feyerabend, der in sehr realistischer Weise einerseits Wissenschaft als ein recht paradoxes System darzustellen vermochte, andererseits meinte, der Erfolg der Wissenschaft bestehe darin, großteils triviale Probleme zu lösen und eben das von Kuhn formulierte Paradigma notorisch zu bestätigen.[3] Einen Fortschritt der Wissenschaft, den Karl Popper mittels seiner spezifischen Methode einer Logik der Forschung[4] im Falsifikationismus zu festigen hoffte, schien es hingegen für Paul Feyerabend keineswegs so eindeutig zu geben, es sei denn, man verbucht die darin enthaltenen Irrtümer als Beitrag zur Ironie. Also kommentierte er diese bekannte Wissenschaftstheorie mit einer Nebenbemerkung, dass sie gerade geeignet sei, zwischen empirischer und ironischer Wissenschaft unterscheiden zu können.[5]

Will man die Position von Feyerabend durch Merkmale seiner Persönlichkeit charakterisieren, so wird man wohl seinen Widerspruch gegen die Dogmatik Poppers und gegen die Behauptung Kuhns, ein Paradigma wirke wie ein Denk- und Sprechverbot in der Wissenschaft etwa in folgender Weise beschreiben können: Feyerabend liebte den denkerischen Exzess, die Paradoxie, fand in den Widersprüchen unserer Tradition des Denkens so manchen Sinn verborgen und bevorzugte Gedankenexperimente, in denen er das Mögliche oder Wirkliche mit dem Unmöglichen und Irrealen zu eigenwilligen Kompositionen gestaltete. Darin entwickelte er eine so große Fertigkeit, dass es nicht einmal möglich war, ihn als einen klassischen Skeptiker zu bezeichnen, selbst wenn dieser Satz allein schon paradox ist, denn Skepsis galt seit dem 17. Jahrhundert wiederum als zulässiges Kontrollinstrument und hatte den Naturwissenschaften zu ihrem Höhenflug verholfen. Man musste eben Feyerabend sein, um Skepsis als notwendig zu bezeichnen, aber ihr gleichzeitig zu unterstellen, ein unmögliches Gedankenexperiment zu sein.[6]

Viele Jahre vergingen, bis endlich dieses wilde Denken in ein Buch gefasst wurde – *Wider den Methodenzwang* von 1975. Was seine Kollegen schon früher an verschiedenen Universitäten vernommen hatten, war nun das Dokument seines Denkens geworden, das mit einiger Verve und überraschenden Gedankensprüngen zwischen den unterschiedlichsten Disziplinen nachwies, Erkenntnistheorie würde weder eine zuverlässige Methode produzieren, noch erlaube sie eine feste Logik der Wissenschaften, da sich ja die Fortschritte unseres Wissens nicht nach strikt rationalen Bestimmungen rekonstruieren lassen. Seine Behauptung, Wissenschafter verteidigen ihre Theorien vornehmlich mit subjektiven und irrationalen Beweggründen, kann als Skandal bewertet werden. Heute mag man im Rückblick behaupten können, Feyerabend habe die postmoderne Diskussion eröffnet, was auch immer darunter zu verstehen sei, er habe eine unhaltbare Wissenschaftsgläubigkeit aufgedeckt und ihr den Mangel an gesellschaftlicher Verantwortung nachgewiesen.[7] Er beschreibt damit den einen und nunmehr geläufigen Umstand, der uns gegenwärtig weit vertrauter ist als noch vor etwa dreißig Jahren.

Das war kein Angriff auf Wissenschaft und Wissenschaftstheorie, wie wir ihn aus der feindseligen Ablehnung von Wissenschaft aus dem 19. Jahrhundert kennen, sondern Feyerabend führte aus, dass Erkenntnisfortschritte eine Portion Irrationalität und Subjektivität benötigen. Wie etwa im Inventar einer Sammlung nicht nur Meisterwerke vorhanden sind, nicht nur Großes und Bedeutendes, sondern da und dort in einer Ecke auch zur Überraschung des Neugierigen Zitate von Kitsch und Sentimentalität zu finden sind, so etwa auch in der Wissenschaft. Schon früher hätte man Feyerabend als einen ersten attraktiven „Hippie" des Denkens bezeichnen können, wären diese Haltung und diese Eigenheit nicht in den zahllosen Kopien und Imitaten im Central

Park oder im Englischen Garten zu München um ihre Originalität gebracht worden. Er war es, weil er ein „anarchistisches" Wissenschaftsbekenntnis bevorzugte und in diesem alles verspottete, was wir als Dogma der Seriosität wissenschaftlichen Arbeitens inzwischen internalisiert haben. Der Titel „Anything goes" sagt genug. Mit wüster Freude zog er über den kritischen Rationalismus her und verletzte alle guten Sitten und Konventionen, die im Wissenschaftsdialog ja ohnehin keinen hohen Standard besitzen. Im gleichen Atemzug, nachdem er Karl Popper gleichsam beleidigt hatte, zog er über Thomas Kuhn her, obwohl ihn mit Kuhn weit mehr verbunden hatte. Da Kuhn bekanntlich den Begriff „Normalzustand" für Wissenschaft beanspruchte, der durch Schulen, dogmatisierte Lehrmeinungen und kanonisierte Theorien aber verletzt erscheint, ein Paradigma eben, so fragte Feyerabend, wann denn und wie oft sich Wissenschaften in diesem Idealzustand befanden? Und dieser provokanten Frage schickte er die Verhöhnung nach, dass die Vorstellung Kuhns vom System „Wissenschaft" als soziopolitisches Modell weit besser im organisierten Verbrechen dargestellt werden könne.[8]

Ja, um diese exzentrischen Angriffe zu tolerieren, war jede Auseinandersetzung mit Feyerabend ein Geduldspiel, eine harte Probe für Höflichkeit und Selbstachtung. Es mag erwähnt sein, dass sowohl Lakatos als auch Popper diesem Exzentriker stets korrekt begegneten, ihm die peinlichen Anwürfe nicht anrechneten, sondern sehr wohl die Einwände beachteten, die sie auch ernst nahmen.[9]

Will man diese Ausflüge Feyerabends nachvollziehen, so sind sie am Beispiel der Medizin bestens belegt. Entweder in der „Erkenntnis für freie Menschen" oder in „Anything goes" beschreibt er eindringlich, dass sich die Erfolge moderner Medizin nicht nachweisen ließen, beziehungsweise nicht wirklich eine verbesserte Wirkung im Vergleich mit den Erfolgen von Medizinmännern zeigen. Ein Mediziner fragte mich einmal, ob nicht dieser Paul Feyerabend ein Narr sei - und ein bösartiger dazu? Will man weiterhin die Frage ernsthaft beantworten, so entdeckt man recht schnell, welchen Gedanken Feyerabend damit zum Ausdruck bringen wollte: Er bemerkte, dass Wissenschaft nicht im bekannten Elfenbeinturm verbleibt, keine Kunst ist, die nur Kunst bleibt, sondern Wissenschaft ist in einem sozialen Raum angesiedelt und da ist dann nur schwer zu bestimmen, welcher Art der Segen ist, der auf die Gesellschaft hereinbricht. Er war eigentlich wie ein scholastischer Denker, der noch vom „mehrfachen Schriftsinn" wusste und daher zu der Überzeugung kam, dass die Dimension von Wissenschaft und Denken weit umfassender ist und sich weiterhin ihrer Prolegomena bewusst sein müsste. So ist etwa sein Plädoyer zu verstehen, wenn er für Schulen den Unterricht in Evolutionstheorie und biblischer Schöpfungsgeschichte verlangte. Die unzulässige Einengung des Vernunftraumes erschien ihm unerträglich und sein Bemühen richtete sich dahin, dass wir doch unseren merkwürdig verschlüsselten und in zahllose Labyrinthe ausufernden Denkraum kennen sollten.

Will man einen Vergleich wagen, so war Jahrzehnte zuvor ein Denker in eine ähnliche Konstellation geraten. Hugo Ball war dadaistischer Dichter und zugleich ein eminenter politischer Beobachter und Philosoph.[10] Paul Feyerabend wies vermutlich eine ähnliche Struktur auf. Seine Methode war eigentlich „Dadaismus" und sie erlaubte es ihm, der Wissenschaft die Neigung zur Tyrannei nachzuweisen. Wie in früheren Gedichten sich Wörter zu reimen hatten und somit der Sprache eine Zwangsjacke angemessen wurde, so war sie nur für ein gelungenes Zeugnis gehalten worden, dass unter dieser Voraussetzung von Reim und Versmaß die künstlerisch-dichterische Wahrheit zum Ausdruck gebracht werden könne. Der Anspruch auf Wahrheit in der Wissenschaft schien ähnliche Bedingungen aufzuweisen, selbst wenn dieser von ihr auf Richtigkeiten reduziert wurde. Der Schritt zur Formalisierung verlieh ihr schließlich Macht und Kompetenz, die die Vielfalt des Denkens, die Originalität der Idee und den Reiz der Unterscheidungen von Kulturen zerstörte. Damit hatte er allerdings das strenge Thema der Erkenntnistheorie weit hinter sich gelassen und immer intensiver ist im Werk das Plädoyer für moralischen und politischen Protest zu erkennen. Spätestens an diesem Punkt verließ Feyerabend die Kategorien der Skepsis und des Skeptizismus. Seine Kritiker mussten die Farbe wechseln. Somit schien Feyerabend zum Relativisten geworden zu sein. Diese Position hat er aber gründlich in seinen literarischen Ausflügen korrigiert ...[11]

Vermutlich trieb er seine Position dort auf die Spitze, wo er sich auf den Nationalsozialismus bezieht. Dass er sich nicht einfach dem allgemeinen Urteil anschließt, das wiederzugeben nicht wirklich interessant ist, so darin Selbstgerechtigkeit und Opportunismus enthalten sind, liegt auf der Hand. Anfänglich hatte ihn ja gerade an der Zeitgeschichte irritiert, dass sie dank eines ex-post-Urteils ohne Geschichtstheorie oder Hypothesen auskommen konnte, wenn sie behauptete, Fakten würden für sich sprechen. Dass er in der Analyse der Konsequenz des Nationalsozialismus erkennt, mit 1945 sei das Problem nicht wirklich beendet worden, markiert ihn als Kenner der Latenzzeiten von unseliger Geistes- und Institutionengeschichte. Er meinte recht genau zu wissen, wovon er da sprach. Er war ja nicht nur sozusagen freiwillig an eine Militärakademie 1942 gegangen, sondern zuvor soll er auch Mitglied der Hitler-Jugend gewesen sein. Es war nicht seine Art, diese prekären Lebenserfahrungen zu verdrängen, was auch nicht möglich gewesen ist, erlitt er doch an der Front gegen Russland eine schwere Verwundung, an der er sein ganzes Leben laborierte.

Das Fazit, das Feyerabend hinterließ, schrieb er in den „Irrwegen der Vernunft" nieder, 1989:

Auschwitz ist nicht etwas, das außerhalb des Menschentums liegt; Auschwitz ist etwas sehr Menschliches, und die Haltung, die dazu führt, ist heute lebendiger denn je.

Sie zeigt sich in der Behandlung von Minoritäten in industriellen Demokratien; in der Erziehung zu einem humanitären Standpunkt eingeschlossen, die immer ausschließlich ist ... Die Einstellung wird manifest in der nuklearen Bedrohung, der Existenz von Feindbildern, der Kalkulation der Auswirkungen eines Nuklearkrieges ... Sie zeigt sich an der zunehmenden Ausrottung der Natur und „primitiver" Kulturen, ohne auch nur einen Gedanken zu verschwenden an das seelische Elend, erzeugt durch den letzteren Prozeß. Der Geist von Auschwitz zeigt sich in dem Mangel an Gefühl vieler sogenannter Forscher, die Tiere systematisch quälen, ihre Qualen studieren und dafür noch Preise bekommen. Was mich betrifft, ist der Unterschied zwischen diesen „Wohltätern der Menschheit" und den Henkern von Auschwitz nicht sehr groß ...[12]

Wahrscheinlich hätte es Feyerabend beeindruckt, wenn er die vergleichbaren Ergebnisse der Analysen von Giorgio Agamben hätte lesen können. In ihnen wird ja gezeigt, dass erst mit Auschwitz die bedrückende Form und Wirklichkeit von Biopolitik erreicht wird, die als eine nicht mehr abzuweisende gesellschaftspolitische Herausforderung vor uns ist. In der Biopolitik sammeln sich die von der Modernen uneingelösten Versprechen an, versucht die „Natur in uns" zum Schweigen zu bringen, drängt die bisherigen Grenzen der Natur zurück und verspricht, sie würde eine „rationale" Gesellschaft entwerfen und entwickeln können.[13] Angekündigt hat dieses Projekt bereits Robespierre, der den Optimismus besaß, dass er das in die Politik umsetzen werde, worüber die Philosophen bislang nur spekulieren konnten.

Wir könnten zum Beispiel erkennen, dass die Verbindung zwischen den Menschenrechten und den modernen biopolitischen Bestimmungen von Souveränität und individueller Kompetenz es erlaubt, die Menschen im allgemeinen in aktive und passive zu unterteilen. Das war ja nicht nur den Historikern der französischen Revolution aufgefallen, sondern schon zum damaligen Zeitpunkt bemerkte Sieyès, dass

> die natürlichen und gesellschaftlichen Rechte diejenigen sind, zu deren Wahrung und Entwicklung die Gesellschaft gegründet worden ist, die politischen Rechte dagegen diejenigen, durch die sich die Gesellschaft bildet. Es ist um der Klarheit des Ausdrucks willen besser, die erste Art passive, die zweite aktive Rechte zu nennen ... Alle Einwohner des Landes müssen in ihm die Rechte passiver Bürger besitzen ...; aber ... nicht alle sind aktive.

Selbst wenn die Schlussfolgerungen aus dieser Feststellung hier nun nicht weiter verfolgt werden, so wird man nicht umhin können, dass Feyerabend sehr früh sich dieses gefährlichen Motivs bewusst war und daraus die Erkenntnis gewann, dass es die Wissenschaften sind, die zu solchen Definitionen neigen und sie zugleich zu einem Status der politischen Umsetzbarkeit erheben. Dass es diese kritische Disposition jeweils gegeben hat, beweist nicht allein die politische Analyse der Rassenlehre von Erich Voegelin in den 20er Jahren,[14] sondern zeigt auch, dass vielfach ausgerechnet im Prozess wissenschaftlicher Modernisierung jene Geistesgegenwart verloren ging, die es vermocht hätte, die jeweiligen politischen Konsequenzen aus Wissenschaft rechtzeitig zu bemerken. Für den Soziologen ist zu erkennen, dass Feyerabend genau an diesem Punkt seinen Kritikhorizont besitzt und es ist umso erstaunlicher, dass es ihm gelungen ist, denn in seiner Ausbildung und in seinem Interessenhaushalt „spukte" alles andere herum als Sozialwissenschaft oder politische Analyse.

Er hatte ja mit Gesangs- und Schauspielunterricht begonnen, beschäftigte sich nebenbei mit Astronomie oder las Reichenbachs *Raum und Zeit*. Und dass ihm dann das kurzfristige Studium der Geschichte und hierauf Philosophie an der Universität Wien zu solchen Sprüngen verholfen hätte, ist mehr als zweifelhaft.

So wurde also aus dem „Dadaisten", besser umgekehrt, aus einem leidlich Klavier spielenden Hobbymusiker und Offiziersanwärter wurde ein dadaistischer Denker, der es hervorragend verstand, in dieses überraschende Kompositionsprinzip des Denkens einzutreten und aus Denk- und Wissensobjekten das neue Bild eines „ready made" zu entwerfen, weit beunruhigender, weit bestürzender als es gemeinhin Wissenschafter von ihrer Tätigkeit vermuten.

An dieser Stelle wird man sich aber hüten müssen, nun im Fachjargon der Philosophie fortzusetzen. Man muss sich auch hüten, hier Ausflüge in die Philosophie zu beginnen, denn es existiert ja bereits der professionelle Vorwurf, dass weder Historiker noch Soziologen ernsthaft diese Kompetenz für sich beanspruchen dürfen. Erlaubt ist gerade noch, gemäß der Betrachtung der Eigenart eines Denkers, Schlussfolgerungen zu ziehen, die nicht so sehr als biographische Notizen dienen, weit mehr jenen schon genannten Denkraum abzugehen, um erkennen zu können, welche Perspektiven von Feyerabend für verbindlich gelten können.

Da ist einmal die Bemerkung in Erinnerung zu rufen, die er häufig gemacht haben soll, dass es eben nur 6 Prozent eines Lebensraumes oder einer Lebensspanne sind, die der Philosophie gewidmet werden. Feyerabend, der damit nicht als Kurzzeit-Denker gelten wollte, beabsichtigte hier nur festzuhalten, dass man einerseits lügt, würde man gleichsam länger philosophiert haben, andererseits sei dadurch vor allem der Humor zu kurz gekommen, der unter anderem auch durch Methodik und Verbissenheit aus der Philosophie vertrieben wurde. Gerade in der Erinnerung an Wien meinte er, dass da zwar über Humor nachgedacht, vielleicht auch geschrieben wurde, aber in der zeitgenössischen Philosophie habe er jeden Stellenwert verloren. Feyerabend nahm damit nicht für sich in Anspruch, Philosophie als Humorist betreiben zu wollen, aber jede Art von Fixierungen war ihm grauenhaft erschienen. Die Selbsteinweisung in den humorlosen Denkraum sah er als Ursache dafür an, dass man schließlich nicht mehr in der Lage ist, sich jene Freiheit zu nehmen, die wohl zu den wichtigsten Eigenschaften des Menschen zählt. Und wer es noch genauer wissen wollte, dem erzählte er, dass er eine Stunde mit Musik grundsätzlich einer Stunde Philosophie vorzieht.

Vermutlich ist darin auch sein Versuch anzusiedeln, Kunst und Wissenschaft zu analysieren, um daraus die Aufforderung abzuleiten, Wissenschaft doch als Kunst zu betreiben. Wenn er etwa als profunder Wissenschaftshistoriker nacherzählt, dass das erste – wenn man will – neuzeitliche Experiment von Brunelleschi vorgenommen wurde, nämlich die Identität der Betrachtung des Baptisteriums in Florenz mit dem angefertigten Gemälde gleichen Inhalts festzustellen, was durch Spiegelung dann auch nachweisbar war, so nennt er hierauf die Bedingungen, unter denen die Reihenfolge eines Experiments möglich ist. Hierauf erörtert er die spitzen Einwände von Leonardo, um schließlich Alois Riegl und dessen Werk über die spätrömische Glasindustrie zu analysieren. Was hier nur kurz angedeutet sein soll, so war eben Feyerabend in der Lage, dass es nicht nur einmal diese Verwandtschaft gegeben hat, die in der Beschreibung der artes liberales ja eine eindeutige

Beziehung haben, sondern dass schließlich der Kunstgeschichte das Malheur passierte, ihre Systematik vorerst der Vertreibung ästhetisch-originaler Kreativität zu verdanken. In dem Maße wird Kunstgeschichte Wissenschaft, sobald sie das Authentisch-Künstlerische zu umgehen vermag oder gar wegretuschiert. Diese Beobachtungen machten Feyerabend gegenüber jedem Faktenpositivismus skeptisch und immun, da doch dieser mehrere Dimensionen des Gegenstands vorerst beseitigte. Ergebnis ist dann, wie Feyerabend schreibt: „Erfolge treten ein, nicht weil man sich an die Vernunft gehalten hat, so wie sie in den bereits errungenen Abstraktionen vorlag, sondern weil man vernünftig genug war, unvernünftig vorzugehen."[15]

Hätte zum Beispiel Talcott Parsons diesen Satz gelesen, würde er sich sehr geärgert haben. Er schwärmte von der einen allgemeinen Theorie der Gesellschaft, unverrückbar und fähig zur Erklärung. Und wirklich schien er als Konstrukteur dieser einen allgemeinen Theorie jede Selbstkritik verloren zu haben, da er schon seinen Kindern den Merksatz eingelernt hatte, ihr Vater sei der erste Soziologe.[16]

Man darf annehmen, dass Feyerabend diese Sensibilität für korrekte Nachvollzüge oder Einwände gegen gängige Muster, Interpretationsstile und Theorien seiner Musikalität verdankt. Sie berechtigte ihn auch, etwa vor Franz Schubert zu warnen. In der Ambivalenz dieser Musik spürte er das Problem auf, dass die musikalische Aussage ins Sentimentale gleitet und dadurch jene nationalen Störfälle möglich werden, selbst wenn der Eintritt in eine zauberhafte Tonlandschaft angeboten wird. Gegen Richard Wagner stellte er Offenbach und Schumann. Sein musikalischer Fixstern war Mozart.

Es ist nun schwer zu rekonstruieren, was Feyerabend mit der Musik gemeint haben kann. Vermutlich stellte sie das beste Beispiel dar, wie ein formales Aussagesystem in einer flüssigen Dynamik funktionieren kann und in Parenthese wird er wohl auf die Idee gekommen sein, dass eine ähnliche Relation im Vernunftraum gegeben sein sollte. Wahrscheinlich hätte man da einen Schlüssel in der Hand, mit dem man wenigstens eines der Schlösser zur „Wissenschaft als Kunst" zu öffnen vermag.[17]

Nicht unerwähnt darf bleiben, dass sein Zugang zur Literatur eine vergleichbare Bedeutung besitzt. Natürlich schätzte er Nestroy und Canetti. Vermutlich kam ihm Canettis Roman *Blendung* in den Sinn, da er sich nach seiner schweren Kriegsverletzung gedacht hatte, „in einem Rollstuhl zu sitzen und zwischen den Regalfluchten voller Bücher auf und ab zu fahren. Was für eine herrliche Aussicht!" Dass zu seinen bevorzugten Dichtern auch Bert Brecht zählte, lässt sich denken. Wie er aber seine Sensibilität gegenüber wissenschaftlichen Texten entwickelte, ist ebenfalls auf eine Vermutung zurückzuführen. In Wien erlernte er die Skepsis gegenüber Schulphilosophie in den Gesprächen mit Ingeborg Bachmann und hatte daraus den Schluss gezogen, dass man sich ja hüten solle, bedenkenlos in die Gehege von Kunst und Religion einzubrechen. Vielleicht fand auch er den gängigen Begriff des „Kunst-Schönen", der da stets in der Ästhetik sein Unwesen treibt, unerträglich. In der Zusammenfassung zum Begriff des Stiles reiht er neun Punkte auf, die sich auch der Wissenschaftstheoretiker merken sollte. Auch die Wissenschaften kennen Stile, wonach deren Erfolg gemessen wird und Objektivität der Anschauung und Darstellung bezieht sich allein nur auf den durch die historische Situation vorgegebenen Sinn. Fast könnte man den Eindruck haben, Feyerabend sei ein Phänomenologe geworden.[18]

Ein bei weitem schwierigeres Thema war ihm die Religion. Ja. Religion, was sicherlich ein nicht nur abgenutzter, sondern auch irreführender Begriff ist, hatte im Denken Feyerabends einen sonderbaren Stellenwert. Nicht in dem Sinn, dass er eine katholische Erziehung hinter sich hatte, kurzzeitig Atheist war, dann aber das Phänomen der Religiosität als Bewusstseinslage in der Gesellschaft – frei nach Durkheim – zur Kenntnis nahm, sondern er akzeptierte den Ernst der Frage vieler Menschen und sagte sich: „Weshalb sich also nicht damit befassen?" Das war für ihn eine zentrale Frage um 1990. Darüber wollte er noch eine eigene Abhandlung schreiben, doch war ihm der Tod zuvorgekommen. Und zugleich gab es weit früher die naive Seite, wenn er etwa seinem Gesprächspartner wegen der Behinderung lauthals über die Gasse zurief: „Soll ich nach Lourdes gehen?" Die Antwort „No, wenn du glaubst", amüsierte ihn, denn in ihr sah er die spannende Ambivalenz des Wortes „glauben".[19] Diese Frage schien ihn denn doch zu beschäftigen, denn noch 1991 hatte er in Wien einen Vortrag über die „Grenzen des Wissens" gehalten. Dabei war dieses Thema nicht innerhalb neuzeitlicher – no sagen wir – Orthodoxie angesiedelt, sondern seine Dissertation hatte sich ja mit der antiken Atomistik befasst – bei Demokrit, Leukippos und Epikurs Kommentaren, also war er gebildet genug, diese Frage aus der antiken Tradition und wegen der Dauerhaftigkeit des Themas zu untersuchen.

Dennoch sollte man zu seiner Frage zurückkehren, die er da so leichtfertig über die Gasse rief. Bei Feyerabend sind alle diese Sätze konkrete Hinweise. Sie sind vielleicht im ersten Hinhören aus Koketterie geformt worden, sogar aus versteckter Eitelkeit, wechseln aber sofort ihre Qualität, tauschen noch im Aussprechen ihren Inhalt aus und berühren unmittelbar eine sehr konkrete Fragestellung. Diese Konkretheit zeigte sich zumeist in der Gestalt von Ambivalenzen und Variationen, was aus seiner Überzeugung abgeleitet werden kann, dass es nicht nur einen Mittelpunkt geben kann. Die Bestimmung eines Mittelpunktes wäre eine unzulässige Verkürzung der Erscheinungsbilder geometrischer Körper. Sein Motiv war stets, sich grundsätzlich seinem Programm verpflichtet zu fühlen, nämlich jede Art von Demarkationslinie des Denkens zu übertreten. Mit Leidenschaft „outete" er sich als Feind jeder Dogmatisierung, weshalb er auch den Kampf Galileis gegen das Heilige Officium untersucht hatte.

Alle diese Splitter würden Feyerabend vornehmlich als enfant terrible charakterisieren. Waren es aber wirklich nur diese Funktion und bewusst zelebrierten Eigenschaften gewesen, die Wissenschaftstheorie zu zwingen, Besseres vorzulegen? War es sein Ziel gewesen, ausschließlich vor den Gefahren der Wissenschaften zu warnen? War er gar ein Feind von Wissenschaft, wenn er ihren gesellschaftlichen Vorrang als absurd darstellte?

Sein Ziel ist vielleicht in zwei Punkten zusammen zu fassen: Es war seine Sache, eine Realgeschichte des Empirismus zu schreiben, natürlich nicht nach physikalischem Muster. Das zweite Ziel war, dass er den Wahnsinn des zu späten Hinterfragens bekämpfte und darin für unser Denken eine mehr als nur hemmende Wirkung feststellte. Alle diese wilden und oft überstürzt erscheinenden Ausflüge ins Denken waren von seinem pantomimischen Auftreten begleitet, wozu er das Gesicht eines knorrigen Kobolds aufsetzte und sein stimmliches Repertoire voll nützte.[20] Bald spottete er, wetterte oder schmeichelte er und flüsterte. Die Hände wirbelten durch die Luft und er gab seine Kenntnisse wie ein Wasserfall preis. Dennoch nannte er sich faul und zuweilen warf er sich vor, ein Großmaul zu sein.

Die beiden Ziele hatte er immerhin erreicht. Er hatte aber auch ein weiteres Ziel erreicht: Er war in eindrucksvoller Weise für die Freiheit des Denkens eingetreten, nicht für jenes der Experten und Spezialisten. Er verkündete mit aller Leidenschaft, diese Fähigkeit dazu besitzt nicht nur jeder Mensch, sondern dazu hat er auch jede nur denkbare Kompetenz. In den Wissenschaften ist ein vergleichbares Plädoyer seit langem nicht mehr vernommen worden. So plausibel diese Verteidigungsrede im Moment auch erscheint, so muss doch angemerkt werden, dass in unserer Fachliteratur über den Wandel der Bildungs- zur Wissensgesellschaft der Name Feyerabend großteils unerwähnt bleibt. So erweist sich gelehrte Ignoranz als die gelungenste und heimtückische Waffe von Wissenschaft, um ihre Position zu verteidigen. In ironischer Weise wird Feyerabend gerade noch dort zitiert, wo immerhin die Frage nach der Vermittlung von Wissen beispielsweise auch „basisdemokratisch" entschieden werden könnte, eine Variante, die wie eine Groteske anmutet, aber damit die Rolle und Bedeutung der anderen administrativ-institutionellen und somit legitim erscheinenden Instanzenzüge für geeigneter hält. Die Ansiedlung von Feyerabend im Absurden ist der eine Teil der Heimtücke, der andere ist noch boshafter: Verschwiegen wird in diesem Zusammenhang, womit Feyerabend seine Einwände gegen Wissen und Wissenschaft begründete. Verschwiegen wird, dass unsere fest gefügte Form des Wissens von problematischen Selektionen abhängt, die eben keine bessere Legitimation beanspruchen können als etwa ein Volksentscheid darüber, welches Wissen zu verbreiten sei. Mag es auch hier eine bewusst unternommene Kontrastierung sein, so soll sie uns sehr deutlich vor Augen führen, dass Feyerabend fast vergeblich gelebt haben könnte. Wir haben uns doch einem Expertenwissen ausgeliefert, das sich anmaßt, unseren Alltag zu bestimmen. Mit Feyerabend kann man formulieren, dass derjenige als Experte gilt, der weit mehr in seine gesellschaftliche Reputation investierte als in die Beibehaltung des sich stets verändernden Standards des Wissens, weshalb der Experte weit mehr seinen Ruf repräsentiert als er tatsächlich ein Fachmann geblieben ist.[21]

Anmerkungen

1. Vgl. insgesamt Friedrich Stadler (Hg.), Kontinuität und Bruch, 1938–1945–1955, Beiträge zur österreichischen Kultur- und Wissenschaftsgeschichte (Wien 1988).
2. Darüber berichtete mir mündlich Michael Benedikt.
3. Gewiss ist diese Position von Thomas Kuhn erst später formuliert worden, allein Feyerabend hatte seine Skepsis darüber gleichsam in Vorwegnahme der Darstellung früh geäußert. Vgl. eben Thomas Kuhn, *Die Struktur wissenschaftlicher Revolutionen* (Frankfurt 1967). Eine eingehende Analyse dieser kritischen Wissenschaftstheorie bietet Paul Hoyningen-Huene, „Kuhn: Paradigmenwechsel", in: Michael Fischer et al., *Paradigmen, Salzburger Schriften zur Rechts-, Staats- und Sozialphilosophie* (München 1997).
4. Erstmals in Wien 1935 publiziert.
5. Dies kommt sehr gut im Briefwechsel mit Hans Albert zum Ausdruck. Vgl. *Paul Feyerabend – Hans Albert, Briefwechsel,* Hrsg. Wilhelm Baum (Frankfurt 1997).
6. Dazu Reinhold Knoll, „Paradigm lost", in: Michael Fischer et al., *Paradigmen,* a.a.O.
7. Diese Position mit allen politischen Konsequenzen erörterte Feyerabend auch in: *Erkenntnis für freie Menschen,* Veränderte Ausgabe (Frankfurt 1980) S. 209ff.
8. Diese unorthodoxen Beurteilungen können in gesammelter Form gelesen werden in: Paul Feyerabend, *Wissenschaft als Kunst* (Frankfurt 1984) Kapitel: „Erkenntnisse zum Überleben", S. 135ff.

9. Diese rege Debatte ohne Verlust an Selbstachtung und gegenseitiger Wertschätzung dokumentierte Wilhelm Baum in: *Paul Feyerabend – Hans Albert, Briefwechsel,* a.a.O.
10. Vgl. dazu Cornelius Zehetner, *Hugo Ball, Portrait einer Philosophie* (Wien 2000).
11. Paul Feyerabend, *Die Torheit der Philosophen. Dialoge über die Erkenntnis* (Frankfurt 1997).
12. Paul Feyerabend, *Irrwege der Vernunft* (Frankfurt 1989) S. 25.
13. Giorgio Agamben, *Homo sacer. Die souveräne Macht und das nackte Leben* (Frankfurt 2002) bzw. ders., *Was von Auschwitz bleibt* (Frankfurt 2003).
14. Eric Voegelin, *Die Rassenidee in der Geistesgeschichte* (Berlin 1933).
15. So etwa wieder in: *Wissenschaft als Kunst,* a.a.O., S.139.
16. So schrieb Talcott Parsons in: „The Prospects of Sociological Theory", in: ders., *Essays in Sociological Theory* (Glencoe 1958) S. 348f.
17. Die Hinweise auf das Verständnis für Musik verdanke ich Michael Benedikt.
18. *Wissenschaft als Kunst,* a.a.O., S. 76ff.
19. So berichtete Michael Benedikt über seine Gespräche mit Paul Feyerabend.
20. Diese Charakterisierung ist recht gut bei der Lektüre seiner Trentiner Vorlesungen nachzuvollziehen: Vgl. Paul Feyerabend, *Widerstreit und Harmonie* (Wien1998).
21. Diese Skepsis wiederholt auch Michel Certeau, *Kunst des Handelns* (Berlin 1988).

HANS SLUGA

DER ERKENNTNISTHEORETISCHE ANARCHISMUS.
PAUL FEYERABEND IN BERKELEY

Als ich in den frühen siebziger Jahren nach Berkeley kam, da stand Paul Feyerabend gerade auf dem Höhepunkt seiner akademischen Popularität. Jedes Semester waren bei ihm viele hunderte von Studenten eingeschrieben. Bald fand ich allerdings heraus, dass er jedem Studenten schon in der ersten Vorlesungsstunde eine Eins für den Kursus versprach. Er fügte dabei noch hinzu, dass es bei ihm natürlich keinerlei Prüfungen oder Hausarbeiten gäbe. Man bekam seine Eins selbst in der Tat, ohne jemals in der Vorlesung gewesen zu sein. Die Einschreibung im Kursus allein genügte. So etwas hatte sich bald auf dem Campus herumgesprochen und wurde insbesondere in unserer Sportabteilung den Studenten zugeflüstert. Als Resultat waren zum Schluss Dutzende von Fussballspielern und anderen Berufssportlern in Feyerabends Kursus eingetragen, um auf diese Art schnell und billig ihren akademischen Notendurchschnitt zu verbessern. Ich erinnere mich auch noch an den Augenblick, an dem unsere Universitätsverwaltung von dieser Sache Wind bekam und Feyerabend zwang, zumindest eine Abschlussprüfung für seinen Kursus abzuhalten. Im nächsten Semester händigte Feyerabend den Studenten zu Beginn der Prüfungsstunde ein Blatt aus, auf dem in grossen Buchstaben feierlich das Wort „Abschlussprüfung" stand und darunter hieß es einfach: „Erzähle mir deinen Lieblingswitz." Jeder Witz, auch selbst der dümmste, wurde dann mit der Note Eins belohnt. Natürlich hat unsere Verwaltung auch diesem Verfahren bald ein Ende gemacht.

Feyerabend wusste wohl, dass viele Studenten überhaupt kein Interesse am Inhalt seiner Vorlesungen hatten und sich nur wegen der einfach zu erhaltenden Einsen bei ihm einschrieben. Er hatte aber seine Gründe für dieses unkonventionelle Vorgehen, wie er mir damals erklärte, und einer davon war erstaunlich konservativen Ursprungs. Er glaubte nämlich zutiefst an das alte Humboldtsche Erziehungsideal der akademischen Freiheit. Diese Einstellung kam auch immer wieder in unserer Fakultätsversammlung zum Ausdruck, wo Feyerabend zur Konsternation meiner Kollegen vehement gegen jede Regulierung des Studienganges argumentierte. Es war allen klar, dass er wenig Sympathie für das verschulte amerikanische Universitätssystem mit seinen fortwährenden Prüfungen, mit seiner Anhäufung von Noten und seiner fetischistischen Anbetung des Notendurchschnitts hatte. Er selbst war in Österreich durch ein freies, Humboldtsches Studium gegangen und konnte sich nicht vorstellen, dass man auf andere Art zu unabhängigem Denken kommen könnte. Er war auch zutiefst überzeugt, wie er mir gegenüber immer wieder betonte, dass er selbst es ohne diese Freiheit nie zu etwas gebracht hätte und er wollte es daher auch den Studenten überlassen, ob und wann und wie sie studierten.

Diese Einstellung erklärt auch, warum er in seinen Vorlesungen so oft auf John Stuart Mills Aufsatz *Über die Freiheit* zu sprechen kam. Mit Mill glaubte er nämlich nicht nur an die überragende Bedeutung der individuellen Freiheit sondern auch an den unabdingbaren Wert einer freien Gesellschaft. Von diesem Einfluss leitet sich auch der Titel von Feyerabends späteren Buch über *Die Wissenschaft in einer freien Gesellschaft* her, in dem er argumentierte, dass auch die Wissenschaften nach ihrem Beitrag zur Bildung und Bewahrung einer freien Gesellschaft beurteilt werden müssen. Mill verteidigte nun ein solches Freiheitsverständnis in seinem Aufsatz wiederum mit Hilfe des Humboldtschen Menschenbegriffs. Nach Humboldt braucht der Mensch Freiheit zur Selbstformierung. Humboldt sah menschliche Individualität also als einen unbedingten Wert an, „as being of intrinsic value," wie Mill zustimmend sagte. Feyerabend interessierte sich nun in seinen Vorlesungen wie auch in seinen Schriften insbesondere für die erkenntnistheoretischen Folgerungen, die Mill aus diesem Humboldtschen Weltbild zog. Mill hatte in diesem Zusammenhange im ersten Kapitel seiner Abhandlung geschrieben, dass wir andere Überzeugungen tolerieren sollten, erstens weil wir uns nie der eigenen hundertprozentig gewiss sein können und zum zweiten weil wir sie erst dadurch richtig verstehen, dass wir sie mit anderen, gegensätzlichen Überzeugungen konfrontieren. Erst im Wechselspiel zwischen verschiedenen Ansichten erhalten unsere eigenen Ansichten, wie Mill es sah, ihr eigentliches, individuelles Profil. Dies war für ihn aber nun selbst wieder eine Konsequenz des Humboldtschen Menschenbegriffs. Ebenso wie menschliche Individualität nur im freien Zwischenspiel zwischen Menschen zustande komme, so argumentierte er, könne die Bestimmtheit unserer Überzeugungen nur im freien Wechselspiel verschiedener Überzeugungen erreicht werden. Feyerabend sprach damals von Mills „wahrhaft humanitärer Verteidigung" dieser Auffassung. (1975, p. 48)

Ich glaube, dass Feyerabends Abhängigkeit von Humboldt und Mill, also den zwei Begründern des modern Liberalismus, nicht nur für seine erkenntnistheoretischen Überlegungen sondern ebenso für seine politischen und sozialen Ansichten bedeutsam war. Insgesamt waren diese nämlich liberal oder selbst liberalistisch, voluntaristisch, und individualistisch ausgerichtet. Das machte er 1978 in *Science in a Free Society* klar, wo er von der „freien Gesellschaft" als einer Gesellschaft sprach, „in der alle Traditionen gleiche Rechte haben und gleichen Zugang zu den Machtzentren" (1978, S. 9) und als einer Gesellschaft, die sich entwickelt „wenn Menschen, die Einzelprobleme im Geiste der Zusammenarbeit lösen, schützende Strukturen schaffen." (1978, S. 30) Da auch die Wissenschaft dieser freien Gesellschaft dienen soll, müssen wir uns immer wieder fragen, ob sie „die Seelen unbeschädigt lässt." (1978, S. 84) oder ob sie Schranken errichtet gegen das, „was Menschen hätten sein können", ob sie „unsere Humanität vermindert" oder es uns erlaubt „frei zu sein". (1970, S. 91) Man muss hinzufügen, dass Feyerabend wenig tat, um diese hier genannten moralischen und politischen Begriffe auszufüllen, aber man darf nicht im Zweifel sein, dass sie allesamt liberalen Ursprungs waren. Er wollte aber seine Auffassungen nicht unbedingt als solche gekennzeichnet sehen, denn der Liberalismus galt im akademischen Milieu der späten sechziger und frühen siebziger Jahre als die reaktionäre Ideologie der herrschenden Klasse, gegen die man damals in den Universitäten so verbissen kämpfte. Feyerabend schrieb 1970 also defensiv: „Viele Leute sind heute geneigt, Mill einen Liberalen zu nennen und ihn wegen der von ihnen wahrgenommenen Schwächen des liberalen Credos abzuweisen. Das ist etwas ungerecht, denn Mill ist sehr von dem verschieden, was wir heute ,Liberalismus' nennen. Er ist in vieler Hinsicht ein *Radikaler*. Selbst als Radikaler ragt er jedoch wegen seiner Rationalität und Humanität hervor." (1970, p. 108) Radikalität, Rationalität (!) und Humanität sollten also die Schlagworte sein, nach denen sowohl die Gesellschaft wie auch die Wissenschaft beurteilt werden müssen.

Feyerabend ging nun allerdings schon 1970 weit über die liberale Position durch seine tiefe Skepsis gegenüber allen gesellschaftlichen Institutionen hinaus. Typisch ist, wenn er zu der Zeit schrieb, dass „alle lang anhaltende Stabilität ... rein und einfach, ein Zeichen von Versagen ist." (1970, S. 30) Auf Grund dieser Haltung hatte sich seine liberale Erziehungsphilosophie in dieser Periode, wie schon beschrieben, zu einer radikal antagonistischen Haltung gegen die akademischen Institutionen verschärft und zugleich hatte sich seine liberale, von Mill geprägte Erkenntnistheorie zu einem erkenntnistheoretischen Anarchismus fortentwickelt. Eine solche Fortbildung ist natürlich nicht ganz unerklärlich, weil der Anarchismus ja in gewisser Weise eine Radikalisierung des Liberalismus darstellt – und zwar eine linke Variante, im Gegensatz zu der rechten, die wir unter dem Namen „Libertarianismus" kennen. Dieser Anarchismus war nun in den frühen siebziger Jahren in akademischen Zirkeln weit verbreitet und er war die treibende Kraft hinter einer langen Reihe von akademischen Protestaktionen, die zwar oft verschiedene Anlässe hatten, deren eigentliches Ziel es aber stets war, die bestehenden Institutionen zu unterminieren. Man erinnere sich, zum Beispiel, daran, wie die Tochter von Jacques Lacan, die zu der Zeit in Vincennes Philosophie unterrichtete, regelmäßig Diplome in Philosophie an die verdutzten Passagiere der Pariser Metro verteilte. Es war genau in dieser Epoche, so erinnere mich noch sehr deutlich, dass Feyerabend einmal ernsthaft auf einer Fakultätssitzung darlegte, wie wir einfach auf der Telegraph Avenue, wo die Studenten, Hippies, und alle möglichen kulturfeindlichen (*countercultural*) Elemente verkehrten, nach einem neuen Assistenzprofessor für unser Institut suchen sollten. Solche Worte machten ihn bei meinen älteren Kollegen zwar schnell unmöglich, brachten ihm aber auch, wenn sie kolportiert wurden, den Applaus unserer Studenten. Feyerabends subversives Vorgehen gegen die Strukturen und Anforderungen des amerikanischen Universitätssystems passte, in anderen Worten, genau in den historisch-politischen Zusammenhang der Periode und musste ihn unweigerlich zum Helden der ähnlich subversiv eingestellten Studenten machen.

Man sieht an diesen Einzelheiten auch, dass es Feyerabend in dieser Zeit in seinem Denken und vorzüglich in seinem Unterricht um viel mehr als Wissensschaftstheorie ging. In unserem Institut galt er allerdings offiziell immer noch als Vertreter dieses philosophischen Spezialfaches, für das er ja in Berkeley eingestellt worden war. Aber wenn er jetzt in seinen Vorlesungen von den Naturwissenschaften und der Wissenschaftstheorie sprach, so tat er es in einer Weise und mit politischen und erzieherischen Intentionen, die meinen Kollegen absolut fremd waren. Feyerabend hatte daher bald keinen leichten Stand mehr in unserem Institut, aber das machte ihm, soweit ich sagen kann, gar nichts aus. Er lachte vielmehr über die Situation und verspottete unsere Kollegen gelegentlich im privaten Gespräch. Die akademischen Philosophen, sagte er mir einmal im Scherz, sind wie Nagetiere. Sie nagen immer weiter am selben Material und man muss sie, wenn nötig, dazu zwingen, von dem alten, toten Holz abzulassen und sich neuen Dingen zuzuwenden. 1970 fasste er seine pädagogischen Absichten in einem Satz von Michael Bakunin zuammen: „Lasst die Menschen sich emanzipieren und sie werden sich aus eigenem Antrieb selbst erziehen." (1970, S. 19) Eine solche Erziehung zur Selbsterziehung sollte den Studenten nach Feyerabend „Daumenregeln, brauchbare Hinweise, und heuristische Vorschläge statt allgemeiner Gesetze" geben, wie er auch schrieb.

Und sie wird diese Hinweise und Vorschläge mit geschichtlichen Episoden in Zusammenhang bringen, sodaß man im Einzelnen sehen kann, wie manche davon manche Leute in manchen Situation zu Erfolg gebracht haben. Sie wird die Vorstellungskraft des Studenten entwickeln ohne ihn jedoch mit präzisen Vorschriften und Prozeduren auszurüsten. Sie wird eher eine Sammlung von Geschichten sein als eine Theorie im eigentlichen Sinne und sie wird ein gutes Stück zielloses Geschwätz enthalten, aus dem sich jeder herausnehmen kann, was zu seinen eigenen Absichten passt. (1970, S. 18-19)

Es waren genau diese philosophischen Voraussetzungen, die einerseits hinter Feyerabends Notenfreizügigkeit steckten und andererseits auch hinter seinem exzentrischen Verhalten gegenüber den Kollegen.

Eines seiner Ziele war dabei die Delegitimisierung des philosophischen Kanons, der philosophisch tolerierten Überzeugungen, und der akademisch geduldeten Theorien und Praktiken. Wenn er, zum Beispiel, mit seinen Studenten den platonischen *Theaetet* besprach, dann wollte er den protagoräisch-heraklitischen Relativismus gegen die platonische Überzeugung von der Objektivität des Wissens verteidigen. Er wollte damit eine Aufwertung der Sophisten gegen Platon und das philosophische Establishment erreichen und zugleich den philosophischen Glauben an die Objektivität in Frage stellen. Es war auch diese subversive Haltung gegenüber der Tradition, und nicht nur die Notenfreiheit, die damals Studenten zu ihm lockte. Er wollte nämlich seine Studenten dazu bringen, auf radikal neuen Wegen zu denken und er wollte ihnen dabei auf den Weg helfen, indem er ihnen zeigte, dass es mehr als ein mögliches Denksystem gibt. Er gebrauchte zur Erläuterung dieses Punktes alle möglichen Illustrationen, wie zum Beispiel Bruno Snells Gegenüberstellung des frühgriechischen parataktischen Denkens und des späteren, logisch-philosophischen und hypotaktischen Denkens. Er sprach in seinen Vorlesungen auch gerne vom Gegensatz, und in der Tat von der Inkommensurabilität des aristotelischen und des modernen physikalischen Weltbildes. Er sprach vom Gegensatz zwischen dem klassisch Newtonischen und dem relativistischen, quantentheoretischen Bild der physikalischen Wirklichkeit. Und er sprach auch davon, dass es neben der wissenschaftlichen Weltbetrachtung vielerlei nichtwissenschaftliche Weltbilder gibt wie die Astrologie, den Zauber- und Aberglauben und die vielen Weltansichten anderer Kulturen. Er interessierte sich in diesem Zusammenhang insbesondere für den anthropologischen Relativismus und das brachte ihn wiederum dazu, den Studenten in anregender Weise, zum Beispiel, von Ruth Benedicts *Patterns of Culture* und von Carlos Castanedas *Teachings of Don Juan* zu berichten. Er konnte jedenfalls, wie ich aus eigener Erfahrung weiss, von allen diesen Dingen in ungeheuer spannender Weise erzählen, denn er war außergewöhnlich belesen und sprach frei und lebendig, indem er die Krücke, die er für das im Krieg lädierte Bein brauchte, als Lehrinstrument, Waffe und Zeigestock benutzte. Später schrieb er einmal von seiner Lehrpraxis: „Von 1958 an war ich Professor der Philosophie an der Universität von Kalifornien in Berkeley. Meine Aufgabe war es, die Erziehungspolitik des kalifornischen Staates durchzuführen ... Ich war mir aber kaum dieser Aufgabe bewußt und hätte sie auch nicht sehr ernst genommen, wenn ich informiert gewesen wäre." (1978, S. 118) Statt zu lehren „was nach dem Beschluss einer kleinen Gruppe weisser Intellektueller Wissen war," hätte er den Studenten einfach erzählt, was er so gelernt hätte, wobei er das Material in einer Weise arrangiert hätte, die ihm „plausibel und interessant" erschienen sei.

Im Rückblick bin ich mir allerdings gar nicht mehr so klar, ob diese Situation für Feyerabends pädagogisches Wirken gut gewesen ist. Es schien mir, zum Beispiel, fraglich, ob er die Gedankengänge seiner Studenten überhaupt kannte oder sich um sie kümmerte. So bestand er den Studenten gegenüber darauf, dass es überhaupt nicht auf Lesewissen ankommt, während er selbst ein dedizierter Leser war, der alles, von antiker Philosophie bis zu den Biographien von Marilyn Monroe und Muhammed Ali, verschlang. Er sah vielleicht gar nicht, dass seine Studenten keinen Ansporn zum Nicht-lesen brauchten. Er sagte den Studenten auch, dass Hexerei genau so gut sei wie die Naturwissenschaft und die hörten dann nicht mehr, wie er hinzufügte, dass er selbst aus diesen und jenen Gründen mehr für Wissenschaft als für Hexerei übrig hätte.

Er verkündete den Studenten sein berühmtes „Anything goes!", machte ihnen aber nicht die von ihm selbst formulierten Qualifikationen dieses Mottos klar. In dieser Periode der „Counterculture" waren Astrologie, Zauberei, andere Kulturen, Gedankenfreiheit, aber auch Aberglaube jeglicher Art nun einmal unter Studenten beliebt, und das akademische und philosophische Establishment mit seinen rationalen und wissenschaftlichen Überzeugungen und Anforderungen unbeliebt. Feyerabend wurde daher von vielen Studenten als ein Guru dieses neuen oppositionellen Denkens angesehen, was er vielleicht nicht bewusst wollte, was ihm aber auch andererseits keineswegs unbequem war. Dieses überhitzte und anarchische Klima brachte ihn nun dazu, seine ohnehin provozierenden Darlegungen immer weiter hochzuschrauben und mit jeder Wiederholung immer weiter zu vereinfachen und zu radikalisieren. In seiner Autobiographie beschreibt Feyerabend, wie er zu dieser Zeit die Erstfassung und die etwas spätere Buchfassung von *Wider den Methodenzwang* geschrieben habe und zwar als „eine Kollage" seiner früheren Analysen und Überlegungen. „Ich arrangierte sie in einer angemessenen Weise, fügte Übergänge hinzu, ersetzte moderate Passagen durch mehr provozierende, und nannte das Resultat ‚Anarchismus'. Ich liebte es, die Menschen zu schockieren." (1995, S. 142) Feyerabend ist in gewisser Weise über seinen erkenntnistheoretischen Anarchismus, dieses Produkt seiner turbulenten Lehrzeit in Berkeley, nie hinausgekommen. Am Ende seines Lebens schrieb er: „Heutezutage bin ich überzeugt, daß dieser ‚Anarchismus' mehr als Rhetorik ist. ... Die Wissenschaftler haben immer in loser und etwas opportunistischer Weise in ihren Forschungen gehandelt ... Es ist angenehm zu sehen, daß einige meiner Lehnstuhl Ideen von Gelehrten beibehalten werden, die in engem Kontakt mit der wissenschaftlichen Praxis arbeiten." (1995, S. 142 u 151.) Das heißt aber nicht, dass sich Feyerabends Denken nach den siebziger Jahren nicht weiter entwickelt hat. So schreibt er in seiner Autobiographie, dass er den kulturellen Relativismus der siebziger Jahre fallen gelassen hat, weil er inzwischen eingesehen habe, „daß jede Kultur potentiell alle Kulturen ist und daß die spezifischen kulturellen Eigenschaften die wechselnden Erscheinungen einer einzigen menschlichen Natur sind." (1995, S. 152) Er erzählte mir später auch einmal, nachdem er seine neue Position an der ETH in Zürich angetreten hatte, dass er in Berkeley zum Schluss geistig hängen geblieben sei, weil er in jedem Semester neue und ziemlich unbelesene Studenten vor sich gehabt hätte, denen er die selben alten Geschichten immer wieder hätte auftischen können. In Zürich dagegen, so fügte er hinzu, säßen nun immer regelmäßig dieselben Studenten und sogar einige seiner Kollegen in seinem Seminar und da müsse er sich anstrengen, neues zu bieten. So sei er, zum Beispiel, auf das genauere Studium der Aristotelischen Physik gekommen und auf die Frage des Verhältnisses von Wissenschaft und Kunst in der Renaissance.

Die anarchischen Zustände im Berkeley der späten sechziger und frühen siebziger Jahre haben jedenfalls unzweifelhaft den Inhalt von Feyerabands Denken beeinflusst und sie liegen in der Tat, wie es scheint, auch der Entwicklung seines erkenntnistheoretischen Anarchismus zu Grunde. Feyerabends Denken wandelte sich jedenfalls in dieser Periode von einer wissenschaftstheoretisch-historischen Ausrichtung zu einer programmatisch-politischen. Als er sich in den sechziger Jahren mit dem Thema der wissenschaftlichen Methodik befasste, da wollte er nämlich anfangs nur zeigen, dass die übliche formallogische Betrachtung dieser Frage zu nichts führt. Wenn die Wissenschaftstheorie rein formal von der Verifikation oder der Falsifikation oder der Bestätigung von Theorien durch die Erfahrung spricht, so waren das nach Feyerabend nur Leerformeln, die nichts über die wirkliche wissenschaftlichen Praxis aussagten. Er argumentierte dagegen, dass wissenschaftliche Methoden sich genauso wie die substantiellen wissenschaftlichen Theorien entwickeln und dass Methoden und Theorien dabei in einer unmittelbaren Wechselbeziehung stehen. Neue Methoden machen neue Theorien möglich, während neue Theorien umgekehrt wiederum neue methodische Verfahren ermöglichen. Diese gegenseitige Abhängigkeit von Methode und Theorie ist allerdings keineswegs logisch determiniert, denn manchmal läuft die Methodik den Theorien voraus und manchmal hat die theoretische Arbeit einen Vorsprung vor der Methodik. Ich erinnere mich noch, wie Feyerabend in Vorlesungen in London in den späten sechziger Jahren diese gegenseitige Abhängigkeit von Theorien und Methoden mit eindringlichen Worten beschrieb. Er sprach damals unter anderem vom Verhältnis von Mathematik und empirischer Wissenschaft und wie die Mathematik manchmal formale Instrumente entwickle, die erst viel später in der empirischen Forschung Anwendung fänden, während bei anderen Gelegenheiten die empirische Forschung den Anstoß zur Entwicklung neuer mathematischer Verfahren gegeben habe.

Feyerabend erörterte in seinen Londoner Vorlesungen, wie so oft, auch die Entwicklung von der aristotelisch-scholastischen zur galileischen Physik. In Übereinstimmung mit Kuhn wollte er zeigen, dass es sich dabei nicht um einen kontinuerlichen Prozess gehandelt habe, der durch neue empirische Erfahrungen notwendig geworden worden wäre. Es ginge in dieser Geschichte vielmehr um einen Methodenwechsel oder, wie Kuhn es ausdrücke, einen Paradigmenwechsel. Die aristotelisch-scholastische Physik habe versucht, das zielbestimmte menschliche Handeln als Modell für alle Naturvorgänge anzusehen; in Galileis Physik sei dagegen die Methode der mathematische Beschreibbarkeit zur Herrschaft gekommen. Die aristotelisch-scholastische Methode hätte somit auch einen direkteren Bezug auf die menschliche Erfahrung gehabt als die neuzeitliche Wissenschaft. Das neue bei Galilei sei eben die platonisierende Ausrichtung auf die Mathematik gewesen und auf die formalen Instrumente, die sie zur Naturbewältigung zur Verfügung stellt. Allgemeiner gesprochen sei die Entwicklung der modernen Naturwissenschaft also eine Wendung zu höherer Abstraktion und damit geradezu weg von der empirisch erfahrenen Welt gewesen. Feyerabend ging aber in diesen Darstellungen noch einen Schritt weiter, indem er hinzufügte, dass Galilei selbst diese methodische Wende noch gar nicht richtig verstanden hätte, und dass er sie auch nicht hätte richtig begründen können, und dass er in seiner Opposition gegen das aristotelische Weltbild opportunistisch alles zur Hand genommen hätte, was ihm methodisch nützlich gewesen sei. Feyerabend schloss daraus, dass es in der Entwicklung der Naturwissenschaften einen ganz normalen und opportunistischen Methodenpluralismus gibt. Diese Tatsache, so glaubte er, bleibe aber oft unerkannt, weil zu vielen Zeiten (wie z.B. in der Periode der aristotelisch-scholastischen Physik) immer nur eine einzige Methode als richtig angesehen werde, die dann oft noch als absolut gültig betrachtet würde. Man müsse also lernen, historisch und interkulturell über das menschliche Wissen nachzudenken, um den richtigen Blick auf die wirkliche Methodenvielfalt der Wissenschaft zu erreichen.

Aber von dieser Position, die Feyerabend solchermaßen in den späten sechziger Jahren vertrat und die sich in vieler Hinsicht an Thomas Kuhns Auffassungen anlehnte, war es noch ein großer Schritt bis zu dem erkenntnistheoretischen Anarchismus, den er in den siebziger Jahren seinen Studenten in Berkeley vortrug. Denn dieser Anarchismus besagte ja nicht nur, dass es in der Tat in jeder Periode der Wissenschaften verschiedene Methoden gibt, sondern er wollte zugleich ein praktisches Programm sein, das zu Vervielfältigung (*proliferation*) von Methoden aufrief. Der Übergang von der ersten, historischen und wissenschaftstheoretischen These zu dieser neuen, programmatisch-politischen ist in der Fassung von *Wider den Methodenzwang* klar angezeigt, die Feyerabend 1970 in *Minnesota Studies* veröffentlicht hat. Da heisst es erst einmal: „Die liberale [d.h. pluralistische] Praxis ... ist nicht nur ein Faktum der Wissenschaftsgeschichte ... Sie ist vernünftig und für das Wachstum des Wissens absolut notwendig." (1970, S. 22) An dieser einführenden Stelle seines Textes argumentiert Feyerabend noch vorsichtig damit, dass es „bei jeder Regel, sei sie noch so ‚grundlegend' immer Umstände gibt, in denen man gut beraten ist, nicht nur die Regel unbeachtet zu lassen sondern ihr Gegenteil anzunehmen." Wenig später im Text macht er aber klar, dass es ihm nicht nur um diese begrenzte Feststellung sondern allgemein um die Rechtfertigung einer unbeschränkten Vervielfältigung von Methoden geht. Er schreibt in diesem Zusammenhang, dass John Stuart Mill zuerst die Idee der Vervielfältigung eingeführt habe und zwar nicht nur als Lösung für spezifisch epistemologische Probleme. „Vervielfältigung wird [von Mill] als Lösung für ein Lebensproblem eingeführt: wie kann ich ein volles Bewußtsein erreichen; wie können wir erfahren, zu was wir zu handeln fähig sind; wie können wir unsere Freiheit vergrößern, sodaß wir darüber entscheiden können, wie wir unsere Talente gebrauchen wollen, anstatt von Gewohnheiten geleitet zu werden?" (1970, S. 28) Zum Abschluss seiner Aufsatzes macht Feyerabend dann noch einmal klar, dass seine Bemerkungen über den Methodenpluralismus einen doppelten Zweck haben. Zum einen den Zweck, das historische Vorgehen der Wissenschaften realistisch zu fassen, zum zweiten, ein praktisches Programm zu verkünden, das sowohl für die Weiterentwicklung des Wissens als für die moralischen und politische Entwicklung des Menschen gelten soll. Er schreibt in dieser doppelten Absicht:

> Die Idee, daß die Wissenschaft nach festen Regeln vorgehen kann oder vorgehen sollte und daß ihre Rationalität im Festhalten an solchen Regeln besteht, ist sowohl unrealistisch wie verwerflich (*vicious*). Sie ist unrealistisch, da sie ein vereinfachtes Bild der menschlichen Talente gibt und der Umstände, die ihre Entwicklung fördert oder verursacht. Sie ist verwerflich, weil der Versuch die Regeln durchzusetzen unzweifelhaft Barrieren errichtet gegen das, was die Menschen sein könnten und unsere Menschlichkeit vermindert während sie unsere professionellen Qualifizierung verbessert. (1979, S. 91)

Man sieht schon an dieser Terminologie, dass es Feyerabend in dieser Bemerkung nicht nur um die erkenntnistheoretische Unterscheidung zwischen realistischen und unrealistischen Beschreibungen der Wissenschaftsgeschichte geht sondern ebenso um den Unterschied zwischen einer humanen und einer verwerflichen Haltung zur Wissenschaft.

Der von Feyerabend propagierte Anarchismus lässt sich allerdings auch mit rein erkenntnistheoretischen Überlegungen nicht plausibel machen. Die von Feyerabend so oft herangezogenen Überlegungen über Galileis methodologischen Opportunismus beweisen, zum Beispiel, höchstens, dass ein solcher Opportunismus zu gewissen Zeiten und in gewissen Situationen nützlich sein kann, nicht, dass er überall und immer richtig ist. Gegen die fest verankerte Methodologie der Aristoteliker und Scholastiker konnte Galilei vielleicht nur seinen Kampf gewinnen, wenn er ihn mit allen verfügbaren Mitteln führte. Methodenvervielfältigung ist in der Tat manchmal, aber nur manchmal, sinnvoll. Wenn sich, wie zum Beispiel im Falle Galileis, die vorherrschende Methode einer Wissenschaft als unfruchtbar erwiesen hat, dann macht es Sinn, nach neuen Untersuchungsmethoden Ausschau zu halten. Auch wenn ein Forschungsgebiet so komplex oder so unüberschaubar ist, dass die vorherrschende Methode den Untersuchungsgegenstand nur teilweise oder ungenügend erfassen kann, dann ist es sinnvoll, neue Methoden ins Auge zu fassen und anzuwenden.

Wir können das vielleicht heute genau in der Philosophie beobachten. Hier hat, zumindet in bestimmten Zirkeln, lange ein rein logisch-analytisches Vorgehen dominiert. Diese Vorherrschaft erklärt sich nun daraus, dass diese Methodik sich, in der Tat, für gewisse Probleme als ungeheuer ertragreich erwiesen hat. Das gilt insbesondere im Bereich der Philosophie der Mathematik und auch im Gebiet der Bedeutungstheorie. Aber selbst in Fragen, welche direkt die Grundlagen der Mathematik und den Begriff der Bedeutung betreffen, hat sich die logisch-analytische Methode nicht in jeder Beziehung bewährt. In diesen beiden Gebieten ergeben sich nämlich auch Fragen, für deren Beantwortung man historische, hermeneutische und phänomenologische Untersuchungsmethoden heranziehen muss. Ähnliches wird deutlich im Gebiet der moralischen und politischen Probleme und noch allgemeiner im Bereich von Fragen, welche sich auf die menschliche Existenz als Ganzes beziehen. In allen diesen Bereichen erweist sich die rein logisch-analytische Methode schnell als unzureichend. Analytische Philosophen haben versucht, auf zwei verschiedenen Pfaden dieser Herausforderung Herr zu werden. Zum einen haben sie versucht, die philosophischen Probleme, soweit es geht, doch noch in das Prokrustesbett ihrer Methodik zu zwingen. Das hat, zum Beispiel, im Bereich der Moral zur zeitweiligen Konzentration auf die Metaethik geführt, wo man, wie es scheint, mit logisch-analytischen Mitteln erfolgreich arbeiten kann. Zum anderen haben die analytischen Philosophen alle Probleme, die sich nicht in einer solchen Weise behandeln lassen, als unphilosophisch beiseite gesetzt. Beide Pfade haben sich aber als nicht ganz erfolgreich erwiesen. Wir sollten daher den Schluss ziehen, dass die logisch-analytische Methode für die Lösung von manchen philosophischen Problemen nützlich ist, aber in anderen Fällen nichts taugt. Das heißt nun nicht, dass wir in solchen Situationen eine spezifische andere Methodik zur Hand nehmen können. Vielmehr scheint es charakteristisch für alle philosophischen Fragen, die unser Menschsein betreffen, dass wir sie nur voll in den Griff bekommen, wenn wir sie von verschiedenen Seiten und in verschiedenen Weisen anpacken. In dieser Situation ist es also durchaus sinnvoll, verschiedene methodische Zugänge auszuprobieren und miteinander konkurrieren zu lassen: ich meine hier phänomenologische, hermeneutische, genealogische, archäologische, historische, dekonstruktive Methoden und was alles weiterhin heute so in der philosophischen Praxis gebraucht wird. Die verschiedenen Probleme in diesen Gebieten formen hier sozusagen einen Körper, den wir von verschiedenen Richtungen her, also mit verschiedenen Methoden durchschneiden müssen. Nur mit Hilfe dieser sich so ergebenden Schnittflächen können wir diesen Körper begrifflich rekonstruieren.

Die Vervielfältigung von Methoden hat aber auch ihren Preis und kann daher nicht überall und in allen Situationen brauchbar sein. Zum ersten muss man nämlich oft lange mit einer bestimmten Methode arbeiten, ehe sie sich als fruchtbar (oder als unfruchtbar) erweist. Methoden sind ja in gewisser Weise überhaupt Langzeitstrategien, und wenn man sich nun erlaubt, freiweg immer neue Methoden zu erfinden, gibt man die möglichen Vorteile solcher Langzeitstrategien auf. In seinen pragmatischen Überlegungen zur Frage, wie man vorgehen soll, wenn man (noch) keine absolut gültige Vorgehensregel besitzt, schreibt Descartes im *Diskurs über die Methode* ganz richtig, dass man in diesem Falle solange wie möglich an einer mehr oder weniger willkürlichen Regel festhalten sollte, weil man am besten aus einem großen Wald herauskommt, indem man immer geradeaus geht. Feyerabends Vervielfältigungsprogramm erkennt zum zweiten nicht, dass die Anwendung einer Vielfalt von Methoden auch Kosten verursacht, zum Beispiel an Denkkraft, an Arbeitskraft, aber auch an Material. Solche Kosten sind zwar in den Geisteswissenschaften und inbesondere in der Philosophie erträglich. Im schlimmsten Falle kommt es dazu, dass ein Philosoph oder eine philosophische Schule sich im Gebrauch von irgendwelchen unfruchtbaren Methoden verschwendet. Aber die Philosophie ist ja überhaupt ein Unternehmen, bei dem das meiste auf dem Schutthaufen der Geschichte oder zumindet auf einem staubigen Bücherregal landet. In den Naturwissenschaften liegt die Sache aber ganz anders. Da verlangt der Einsatz einer bestimmten Untersuchungsmethode oft gewaltige Investititionen an Geisteskraft, Zeit und Material. Das macht die Vervielfältigung von Methoden in vielen Situationen nicht nur unattraktiv, sondern auch oft im Sinne Machs unökonomisch. Wenn Feyerabend von einer solchen Vervielfältigung sprach, dann dachte er natürlich immer an die Naturwissenschaften. Aber es ist möglich, und ich glaube auch eigentlich plausibel, dass seine Lektionen zur Methodenfrage besser auf die Philosophie als auf die Naturwissenschaften passen. Denn die Philosophie ist ja im Gegensatz zur Wissenschaft eigentlich – auch wenn sie das selbst nicht immer erkennt – ein durch und durch anarchisches Unternehmen.

Bibliographie

Paul Feyerabend, "Against Method", *Minnesota Studies in the Philosophy of Science,* Bd. 4, University of Minnesota Press, Minneapolis 1970, 17-130.
Paul Feyerabend, *Against Method. Outline of an anarchistic theory of knowledge,* NLB, London 1975.
Paul Feyerabend, *Science in a Free Society,* NLB, London 1978.
Paul Feyerabend, *Killing Time,* The University of Chicago Press, Chicago und London 1995.

KARL SVOZIL[1]

FEYERABEND AND PHYSICS

Abstract

Feyerabend frequently discussed physics. He also referred to the history of the subject when motivating his philosophy of science. Alas, as some examples show, his understanding of physics remained superficial. Indeed, partly due to the complexity of the formalism which has left many philosophers at a loss, physicists have attempted to develop their own meaning of the quanta. This has stimulated a new kind of empiricism, an *experimental philosophy,* which is plagued by the inevitable interpretation of the raw data, in particular incommensurability. Feyerabend has expressed profound insights into methodological issues related to the progress of physics, a legacy which remains to be implemented in the times to come: the conquest of abundance, the richness of reality, the many worlds which still await discovery, and the vast openness of the physical universe.

Preamble

In the early morning hours before this talk, I had a horrifying dream. I found myself in the position of being expelled from the physics department. I enter it lately, coming home to my institute, either from some mushroom picking or from this conference. The atmosphere is hostile. I walk to my room. The room is occupied with some post-doc students of the department head. The windows which usually overlook the city center are blinded. I am told that the head of the department was trying to reach me the entire day, and that he summons me up on a very grave and serious affair. When I enter his gigantic office, he sits at a huge table. Other very serious members of the institute are gathered as well. They immediately tell me to get seated and listen to the indictments. When I try to recall which scientific crimes I could have possibly committed, I wake up.

In retrospect, I know what crimes I have committed: Long time ago, in almost another life with other persons and other institutions, I have told them "the truth," at least my visions of "the truth." These visions were in many ways totally off mainstream, and I suffered from the disguise of my colleagues. In Berkeley, I had to appeal to the head of the *Lawrence Berkeley Laboratory*'s physics department to get a paper on relativity theory published as an *LBL* preprint [65, 66, 71] which was rejected by Chew on the basis of Stapp's judgment that, if my recollection is correct *"this is not the way to proceed."* Not that I did not also pursue "normal" science, publishable in *Physical Review Letters*, in *Physical Review* or in the *Journal of Mathematical Physics*. But I did also crazy stuff, for which I suffered in my early days. Even later on, when I spoke in a conference organized by the *Institut Wiener Kreis* about the *physics of virtual realities* [68], some of the material contained in my book on *Randomness and Undecidability in Physics* [67], I still remember Professor Flamm shaking his head in disguise, saying, *"Dieser Svozil ist total übergeschnappt"* (in English *"Svozil has gone totally crazy"*), and one of the organizers, Jimmy Schimanovich, later tried to propitiate me with the words, *"but at least you have one advantage over most of the other speakers: you have already prepared your manuscript before your talk!"*

I am deeply thankful to Paul Feyerabend for emphasizing so unequivocally the necessity and the value of original research and the pursuit of "crazy" and unfashionable ideas [19] and methods. I know that many people, including Lakatos, Kuhn, Dyson and others before and after him have expressed this necessity, but never were they so outspoken as Feyerabend. He gave all those talented original undergraduates and young scientists in the wild a clear message which could help to set them free, thereby giving science yet another unexpected turn. In the words of the *Bhagavad Gita,* "go out and conquer yourself a prosperous kingdom! "

1 General attitude

In his autobiography Feyerabend admitted that for his Ph.D. thesis supervised by Hans Thirring[2] he had started working on a problem of classical electrodynamics which he could not solve (p. 85 of Ref. [33]). He then turned to Kraft and to Thirring who accepted a thesis not in physics proper, but in the philosophy of science. Later, Feyerabend wrote several papers [55, 56] on physics-related topics, in particular on the interpretation on quantum mechanics, on classical and on statistical physics, which are reprinted mainly in the first volume of his *Philosophical Papers* [31, 29, 34]. Unlike Popper's attempts to "falsify the Copenhagen interpretation" [57] and argue against the quantum logic introduced by Birkhoff and von Neumann [16], Feyerabend pursued these investigations in a cautious, considerate and self-critical style.

Professor Fischer recalls [36] that the physicists in Berkeley, in particular Karplus [37], generously evaluated Feyerabend to be "merely" two decades behind current research, the average philosopher being at least half-a-century behind. Also, Fischer recalls, Feyerabend was happy with the evaluation, and told him that there was no essential difference between a physicist and a good philosopher, and that Feyerabend considered himself to be too stupid to be a good physicist: "Apart from his stupidity – he assured me – nothing separated him from being a physicist."

I am inclined to agree with this self-evaluation only partially. Feyerabend certainly was very intelligent and a person full of resources. It may well be that he did not want to be bothered with the sometimes tedious task to work out theories formally, or to set up and run experiments.

For whatever reasons, Feyerabend's contributions to physics were minor. In contrast to physics, Feyerabend's contributions and insights into methodological issues are, at least in my opinion, remarkable. The style in which his statements were expressed was provocative; sometimes even bordering to the offensive; always gathering attention and raising eyebrows.

Getting attention was certainly one of his biggest intentions. He did neither succeed as an opera singer, nor at the theater, but certainly at the academic stage. Often the reactions were harsh. In an article published in *Nature* (p. 596 of Ref. [78]), Feyerabend was referred to as "the Salvador Dali of academic philosophy, and currently the worst enemy of science;" a denunciation which deeply saddened him (Chapter 12 of Ref. [33]). I do not think that such a term is justified. Popper with his naive viewpoints and his talk about *"blablabla"* certainly did more harm to science [72] than any other dilettante claiming to know the proceeds of science before; but not Feyerabend. On the contrary I believe that Feyerabend was right in suggesting that input from the outside does science proper good; even if one is not willing to grant that "science has now become as oppressive as the ideologies it had once to fight" [24, 26].

Besides his methodological openness, Feyerabend's lasting message, in my opinion, is the "conquest of abundance," the "richness" of the phenomena around us, and the "vastness" of the territories still awaiting to be discovered. Of course, this message, as many things Feyerabend said, is not entirely new. One finds similarities with Bergson, Broad, in Huxley's *Doors of Perception,* as well as in modern neurophysiologic investigations. But it is still worth stressing that the restricted view of the world in the present scientific perspective is rather a consequence of tradeoffs between comprehensibility and exhaustiveness than a property of nature. We are just at the beginning of the scientific revolutions, and there are numerous challenging and worthwhile tasks out there for the generations to come. The pursuit of science is one of the greatest passions of life, and our capabilities to recognize and manipulate the physical world may only be limited by our phantasy. Maybe one hopefully happy day we will be able to *tune* the world according to our will alone.

2 Tower of Pisa example in "Against Method"

One of the things which Feyerabend discussed in *Against Method* [23] in greater detail is the Tower of Pisa example. It is about an old argument against earth rotation which has been already put forward by Aristotle: A stone from a high tower arrives at the foot of the tower seemingly without any shift relative to the horizontal position of the release point on top of the tower.

Admittedly, Feyerabend had other objectives in mind, in particular some supposed "deceptions" by Galilei, who allegedly "brushed aside" topics seemingly in conflict with his heliocentric approach by maintaining that the phenomena could be correctly described while at the same time "hiding" new "absurd" theoretical assumptions. Feyerabend completely omitted the contemporary physics of the Tower of Pisa example.

Indeed, Galilei seems to have committed himself to the attitude that there should be no shift whatsoever, a wrong conjecture which also seemed to have been accepted by Copernicus. Newton and Hooke investigated this topic more carefully. Indeed, this may have been the starting point of Newton's theory of gravity. Incidentally, also Gauss and Laplace held wrong theoretical opinions on the phenomenon.

After a succession of inconclusive measurements by different researchers, Hall performed experiments in Harvard in 1902 [41, 42]. Due to the admirable effort of the *American Physical Society* to retroscan their entire collection of scholarly articles published in the *Physical Reviews,* Hall's superbly written contributions are easily obtainable. A later review by Armitage [1] which is also cited in *Against Method* states,

... Thus Newton's experimental test for the diurnal rotation of the Earth may be said to have given positive results of the expected order of magnitude, though the persistent occurrence of an unaccountable southward deviation has continued to be a matter for inconclusive speculation.

Despite our present conception of a ferocious earth rotation, which reaches its peak of 464 m/sec or 1670 km/hour at the equator, and which may give rise to measurable effects even if the relative motions are assumed to be small, in his writings Feyerabend never mentioned the contemporary physical situation, in particular the Coriolis force and the Kepler problem. This seems to be characteristic for the attitude of many philosophers of science, as Feyerabend himself polemically notes [24, 26],

... Kuhn encourages people who have no idea why a stone falls to the ground to talk with assurance about scientific method. Now I have no objection to incompetence but I do object when incompetence is accompanied by boredom and self-righteousness. And this is exactly what happens. ...

When one reads these strong words, written in an intellectual climate of the seventies of the past century, one has little doubt that the boldness and self-esteem of such statements provoked antagonism.

Coming back to Tower of Pisa example, some model calculations were done by Martina Jedinger and Iva Brezinova here in Vienna, yielding a latitudinal shift of 9.6 cm towards South and a longitudinal shift of 0.6 cm towards East. Intuitively, the large latitudinal shift could be understood by considering that (air resistance left aside), the falling body remains in a plane spanned by the direction of the gravity pull towards the center of the earth, and by the direction of velocity at its release point. At the same time, the earth, and with it the foot of the tower, revolves around an axis which is currently tilted at 23.5° with respect to the ecliptic axis, the line drawn from the center of the earth and perpendicular to the ecliptic plane; a configuration depicted in Fig. 1.

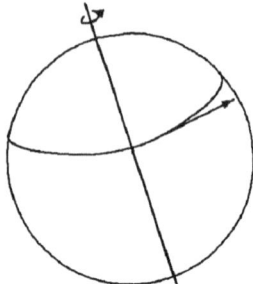

Figure 1: Direction of inertial motion of an object released from a point close to the earth's surface.

In principle, such a setup could even measure the configuration of distant masses by Mach's principle. Recall that, according to Einstein's perception of Mach, the inertial motion of a body should be determined in relation to all other bodies in the universe; in short, "matter there governs inertia here." As the earth's gravity pull is known and the shift of falling bodies is measurable, a reverse computation could yield the inertial motion the distant masses measurable by falling bodies. But this is beyond the scope of this little review.

3 Quantum mechanics

Feyerabend wrote several contributions to the foundational debate in quantum mechanics. They are quite detailed and reflect the ongoing debate at the time they were written, but I failed to find new aspects in them which had a lasting impact on the community. At least Feyerabend was cautious enough not to state any erroneous claims, as Popper did.

3.1 Feyerabend's writings on quantum mechanics

The first volume of the *Philosophical Papers* [31] contains the following five manuscripts on quantum mechanics in consecutive order: *On the quantum theory of measurement* [28], *Professor Bohm's philosophy of nature* [30], *Reichenbach's interpretation of quantum mechanics* [32], *Niels Bohr's world view* [27], and *Hidden variables and the argument of Einstein, Podolsky and Rosen* [25].

In *On the quantum theory of measurement* [28], Feyerabend attempted a reconciliation between the two types of time evolutions in quantum mechanics: the unitary, reversible evolution of the state inbetween measurements, and the irreversible "reduction of the wavepacket," or "collapse of the wave function" – if such notions are appropriate – in a classical measurement device, producing for instance a click in a particle detector. Presently, the situation regarding this issue appears as unsettled as ever, despite some dramatic empirical developments through single quantum experiments; in particular the reconstruction of quantum states after (reversible) "measurements" such as the quantum "eraser" experiments (e.g., Refs. [44, 40]).

Professor Bohm's philosophy of nature [30] is a critical evaluation of Bohm's theory of hidden parameters. Feyerabend expresses his mixed feeling of the book: on the one hand, its approach is fresh and original, on the other hand Feyerabend is reluctant to abandon the traditional Copenhagen interpretation of quantum mechanics.

Niels Bohr's world view [27] starts with a refutation of an erroneous claim by Popper to have falsified the Copenhagen interpretation (see also Ref. [57]). It is also an almost heroic monumental effort to understand that interpretation and its alleged creator, Bohr. The paper contains 101 references.

In *Hidden variables and the argument of Einstein, Podolsky and Rosen* [25], Feyerabend reviews the paper by these three authors [22]. Its first sentence appears to be slightly misleading, at least to me: "Opponents of Bohr's interpretation often refer to an argument by Einstein, Podolsky and Rosen, (EPR) according to which, the formalism of wave mechanics is such that it demands the existence of exact simultaneous values of non-commuting observables."

It is the particular physical setup using two correlated particles which allows the measurement of two non-commuting observables on two particles, one observable per particle. Through counterfactual inference, this property is then ascribed to the partner particle as well, and *vice versa*. In that counterfactual way, one may insist to be able to "measure" two observables which are non-commuting and thus non-comeasurable on a single particle quantum mechanically. So, Einstein, Podolsky and Rosen claim, quantum mechanics is incomplete, since one can measure more than this theory is able to predict. Feyerabend then proceeds to derive consequences of such a hypothetical "more complete" theory, allowing "superstates" through hidden parameters.

Feyerabend's general approach in this debate seems to be dominated by an antagonism against Popper. As Popper favors realism and argues against Bohr's Copenhagen interpretation, Feyerabend objects and argues in Bohr's favor; although cautiously and with many reservations. There seemed to have been even a mini-foundational debate between philosophers of science going on, which developed in parallel to the physical debate, and which was almost totally neglected by the physicists. At least for me, this debate seems to have lead nowhere. But Feyerabend is here in good company with very many physicists and laymen alike.

3.2 Philosophers at a loss to understand the new physics

Recall Feyerabend's statement cited above on people who have no idea why a stone falls to the ground talking with assurance about scientific method; where incompetence is accompanied by boredom and self-righteousness. These are very harsh, critical words which in my opinion characterize Feyerabends (self-) provoking stile. They are, I think, not entirely unjust. Indeed, the main premise in my opinion *is* correct: most philosophers nowadays are at a complete loss of understanding the more recent developments in physics. With *philosophers* I mean everybody with an academic degree after a study mainly concentrating on philosophy, as compared to the natural sciences.

There are great exceptions to the rule, but these are rare and sparse. I certainly do not want to contribute to the ridiculous debate of the natural sciences with the rest of the faculties, sparked by Sokal [60], as I certainly do not want to argue that the natural sciences are immune to fraud, misbehavior, stupidity and deception. All I want to say is that any philosophy of science will be misleading without a proper education in and knowledge of the subject. This is particularly true for the philosophy of science. So, I am afraid, I have to urge philosophers and students of philosophy of science to study mathematics, physics, logic, chemistry, biology and computer science proper. At least the mastering of one of these subjects is necessary in order to be able to comprehend, more so to contribute, to the ongoing debates in these areas.

In the meantime, physicists like myself will go wild and usurp territories which would be better covered by the philosophers, as they have much more background in the historical debates and are less inclined to state ridiculously naive claims on foundational questions such as reality and metaphysics. We desperately need philosophy after all, as we desperately need philosophers! But we need to educate them better in the sciences, if they wish to consider science. And please do not confuse attempts to brainwash people into science proper with concerns of competence.

From these very general remarks, let me now come back to quantum physics, which still remains a very active research area. Almost since its introduction in 1900 it has been the subject of intense philosophical debates, both within the physics community — at that time in central Europe, the physicists, due to the good old Humboldt type curriculum, were much better trained in classical philosophy — and among philosophers of science. Also Feyerabend contributed to this debate, as already mentioned. If one is not willing to digest the volumes of Jammer [47, 48, 49], or the collection of original articles by Wheeler and Zurek [80], one gets a good glimpse of what was and still is going on from Schrödinger's series of three articles on *"Die gegenwärtige Situation in der Quantenmechanik"* [58] (English translation *"The Present Situation In Quantum Mechanics"* [80]). I think that I can safely say that, although "nobody understands quantum mechanics" (cf. Richard Feynman in Ref. [35], p. 129), nobody not able to comprehend these Schrödinger articles should make a public appearance on related topics.

3.3 Old topics in a new terminology: scholasticism, empiricism, realism–idealism

The debate on the foundations of quantum mechanics has taken a somewhat unexpected turn, in particular in recent years, when, due to new experimental techniques, single quanta could be investigated: it turned into controversies with associated long-lasting debates and huge philosophical records. At the same time, some of the old concepts became formalized. In what follows, I take up the task of reviewing some of these issues, being well aware of the risk of being dilettantish.

One of the big issues is related to the realism *versus* idealism debate. Stace [63] characterized realism by the supposition that "some entities sometimes exist without being experienced by any finite mind." He was an outspoken idealist, claiming that

> ... we have not the faintest reason for believing in the existence of inexperienced entities ... [[Realism]] has been adopted ... solely because it simplifies our view of the universe.

In quantum mechanics, this debate may probably be summed up by the term *Bohr–Einstein debate,* with Einstein firmly positioned as a realist. One of the questions concerns the existence of physical properties even in the absence of their direct physical observation. Einstein [22] suggested to do just that, and effectively accept indirectly inferred counterfactuals als *"elements of physical reality."*

A further step was taken by the Swiss mathematician Specker, who, stimulated by the quantum logic developed by Birkhoff and von Neumann [8], pondered about the logic of propositions which are not co-measurable; i.e., not simultaneously measurable [61]. Specker related such structures in quantum physics to Scholasticism, in particular to scholastic speculations about the existence of "infuturabilities" or "counterfactuals." The question is whether or not the omniscience (comprehensive knowledge) of God extends to events which would have occurred if something had happened which did not happen. If so, could all such events be pasted together to form a consistent whole?

Concerns about co-measurability were inevitable because quantum mechanics had introduced new features hitherto hardly heard of in classical physics. Complementarity is a system property first discovered in quantum theory, making it impossible to jointly measure two complementary observables; or at least prohibiting their joint measurement with arbitrary precision. The associated non-commutative operators and the resulting non-distributive propositional structure became facts of everyday professional life for theoreticians and experimentalists alike. Yet, what is hardly noticed even by the specialists is the fact that complementarity and non-distributivity not necessarily implies total abandonment of non-classicality: quasi-classical models such as generalized urn models [81, 82] and finite automata (e.g., Chapter 10 of Ref. [70]) can be isomorphically embedded into Boolean algebras with the help of two-valued probability measures, which abound for such models [73].

In retrospect, that such non-co-measurable propositions exist might have come as no surprise even to the classical mind in retrospect. Indeed, one could take this as a good example for the fact that our phantasy is not good and wild enough to conceive of the many available alternative options we have for almost any given situation.

Bell [6] and others took up an idea expressed in the influential article of Einstein, Podolsky and Rosen [22], to which also Feyerabend referred. The Bell-type inequalities are particular instances of Boole's *"conditions of possible experience"* [10, 9], which are consistency conditions on joint probabilities, for specific physical setups. The Einstein-Podolsky-Rosen argument suggests that, although two non-co-measurable properties (associated with complementary observables) cannot be directly measured at a single quantum, one may infer from certain two-quanta states (satisfying a uniqueness property [75]) one property per quantum and subsequently counterfactually infer the other property from its twin quantum.

In doing so one implicitly assumes that counterfactuals exist: it is assumed that, if the counterfactually inferred property would have been measured — which was not the case — it would have come out in the expected way.

On top of that, the Bell inequalities contain sums of terms (e.g., joint probabilities or expectation values) which could only be measured subsequently (or at least in different experimental setups; one at a time), since they correspond to different parameter settings which, according to quantum complementarity, cannot be measured simultaneously. To the realist assuming that entities exist without being experienced by any finite mind, this is no big deal. Even collecting terms associated with measuring different non-co-measurable setups and summing them up as if they referred to a single quantum is hardly disturbing. (This makes possible a criticism put forward recently [45].)

But the assumption of an "all out" omni-realism may be improper in the quantum domain. Quanta prepared in a specific state in a given experimental context might simply not be capable to "know" their precise states in different contexts [76].

As an analogy, no finite agent such as a computer program can be set up to answer all conceivable questions — it may be at a complete loss at answering some or even most of them. This kind of restricted capabilities appears quite natural and is not very exciting; certainly not as "mind-boggling" or "mystical" as the quantum tales of Bohr. It is a new form of realism, one which is based on the assumption that certain things do not have all conceivable properties we would wish them to have; just a finite number of properties, that is it.

Nevertheless, let me emphasize that the non-classicality of quantum mechanics goes well beyond complementarity. There is a finite constructive proof of the impossibility of value definiteness for quantized systems whose description require Hilbert spaces of dimension higher than two. It turned out that formally there are "not enough" two-valued states to allow a faithful embedding of certain tightly interconnected finite propositional structures into any Boolean algebra. This can be characterized by non-separable or non-unital sets of two-valued states; the strongest result being the non-existence of two-valued states known today as the "Kochen-Specker theorem" [50]. In fact, once the propositional structure has been enumerated explicitly, a proof is technically not very demanding and amounts to coloring (chromaticity) these sets.

To get a taste of the type of argument, Fig. 2 depicts a quantum propositional structure; i.e., a logic, with a non-separable set of two-valued states, such that $P(a)=P(b)$. It is the graph G_3 of Ref. [50] represented by the Greechie orthogonality diagram in Ref. [77].

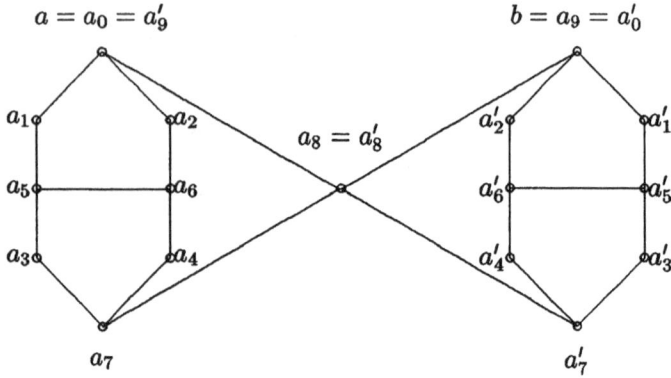

Figure 2: Greechie (orthogonality) diagram [39], consisting of *points* which symbolize observables (representable by the spans of vectors in n-dimensional Hilbert space). All points belonging to a context; i.e., to a maximal set of co-measurable observables (representable as some orthonormal basis of n-dimensional Hilbert space), are connected by *smooth curves*. Two smooth curves may be crossing in common *link observables*. In three dimensions, smooth curves and the associated points stand for tripods. Two valued measures (interpretable as truth assignments) assign exactly one point per context the value 1; the other elements are 0. If $P(a = a_0 = a_0') = 1$ for any two-valued probability measure P, then $P(a_8) = 0$. Furthermore, $P(a_7) = 0$, since by a similar argument $P(a)=1$ implies $P(a_7) = 0$: for, if $P(a_7) = 1$, then $P(a_1) = P(a_2) = P(a_3) = P(a4) = 0$, resulting in the necessity for $P(a_5) = P(a_6) = 1$, which is contradicting the assumption that there can only be one element per context which is 1. Therefore, $P(b = a_9 = a_0') = 1$. Symmetry requires that the reverse implication is also fulfilled, and therefore $P(b) = P(a)$ for every two-valued probability measure P.

4 Physicists at their own: incommensurabilities in experimental philosophy

4.1 A "meaning" of the quantum?

So what have the physicists produced when they were left alone to interpret the "meaning" of the formalism? They have developed a variety of interpretations, the number of which is probably as great as there are physicists and as the number of philosophical concepts of reality. Let us just mention a couple of these interpretations and some of their creators and devotees: the Copenhagen interpretation (Bohr), the many-worlds interpretation (Everett), Bohm's formalism, the consistent histories approach (Griffiths) and finally a "realistic" interpretation (Einstein, De Broglie, Schrödinger). All of these make no empirical difference. Yet they serve as a kind of scaffolding [74]; without it science would be reduced to an automated proof technique, playing havoc without guidance, devoid of any idea of how to proceed with (hopefully) progressive research programs [51].

4.2 Attempts toward a new kind of experimental philosophy

Interpretations aside, it is tempting to speak of a kind of "experimental philosophy," by which ancient questions of philosophy can be decided by empirical techniques; i.e., in the laboratory. This, it seems, is just another kind of extreme empiricism. But beware of incommensurabilities when interpreting the raw data: terms in different theories do not share the same meaning and thus cannot be directly related and compared. Stated differently, incommensurability asserts that there is an inevitable lack of common theoretical as well as of operational terms due to conceptual differences.

Already Hertz [43] spoke of different illusory "images" or "models" (German "*Bilder*") of the mechanical laws, being all in a certain correspondence with the sense data. As examples, he mentioned the concepts of *force*, as expressed in Newtonian physics, opposed to *energy*, expressed in the Hamiltonian formalism. These images are illusions, which can be applied to the particular purpose they were created for, but otherwise cannot be taken for reality. In Hertz's own words (cf. page 1 of Ref. [43]),

> we create illusory inner images, or symbols of the external objects, such that the consequences of these images are always the images of the consequences of the objects represented. ... Besides this, we have no certainty that our conceptions of the objects have anything else in common than this *single* fundamental relation.[3]

4.3 A philosophy from detector clicks?

The raw data *per se* cannot decide ancient questions of philosophy – how could they? Dichotomic events such as clicks in a counter decide just one thing: whether there is or there is not such click. Thus any interpretation of the raw data inevitably introduces theoretical constructions which are in no direct correspondence to the empirical basis, and may change as times and fashions go by. All we can ever observe as sense data are ultimately discrete events such as detector clicks. There is no other empirical basis than such clicks; they are all we have in constructing what we call the world. During that world construction, conventions and theories enter. But conventions and theories are neither evident nor eternal. They just reflect reasonable assumptions. For example, the operationalist Bridgman, in a heroic attempt to base physics on empirical grounds alone, had to introduce more and more theory to make sense of the raw data [11, 12, 13, 14, 15].

The foundational debate in quantum mechanics in general, and the interpretation of the experiments first performed by Aspect, Grangier and Roger [2, 3, 4] on the violation of Bell-type inequalities in particular, demonstrate this quite clearly. The basis of claims of the so-called "quantum non-locality" is the occurrence of pairs of clicks of a specific type in some detectors. Occasionally, a click here and a click there, that is all, isn't it? For special parameter configurations, there are more or less joint clicks than can be expected classically.

Quantum theory gives "meaning" to these (in) frequent occurrences of pairs of clicks; but besides quantum theory, several alternative theories could also explain the clicks. You may not like these theories, since they do not give any new insights and look suspiciously artificial, but that may not be a good criterion to favor one physical theory over the other. Certainly we cannot rule out that, as time goes by and physics progresses, other theories might give even better interpretations of the raw data than quantum mechanics. So, beware of "experimental philosophies" which only superficially seem to be corroborated by the empirical records.

Another example of incommensurability in physics is the interpretation of the Michelson Morley interference experiment, which gave a "null" result for a model of the luminiferous ether drift [53, 59]. At first, they were interpreted by FitzGerald [38, 7], Larmor, Lorentz and others as an indication that the ether may effectively "shrink" objects in very much the same way as we perceive Lorentz contractions nowadays. Later on, Einstein's theory of special relativity [20] was interpreted as proving that an ether "does not exist." Einstein himself [21], as well as for instance Dirac [18] and Bell [5, 7] held much more differentiated opinions on this subject. So, again, there does not seem to be a unique way of interpreting the raw data of the Michelson Morley interference experiment [64].

Let me also mention a somewhat unrelated issue. Some physicists "go wild" and pretend that the transient status of their science reflects final truth of the world; they tell fairy tales about the first three minutes of the Universe, short histories of time and what not. This is good for marketing purposes and sells well. What they do not seem to acknowledge and the public simply does not want to hear is the historic aspect of our findings which makes our present knowledge transient and preliminary.

In this way of thinking, which emphasizes transitions and the continuation of research programs, I tend to agree with Lakatos [51]. Feyerabend's critique of Lakatos is that the latter does not offer a methodology. Yet the same could be said of Feyerabend's methodology [24, 26]. And if openness, or suspended attention as Freud put it, is no methodology, then so be it.

5 Some personal remarks

5.1 Personal encounters

In the spring semester 1983, I attended a course of Feyerabend on the philosophy of science in Berkeley during my stay as a visiting scholar at the Lawrence Berkeley Laboratory and the University. Feyerabend made a sad but rebellious impression, best described by the German word "unerfüllt," whose English translation is "unrealized, unfulfilled." His spirits were strong and he gave a quite good performance.

His audience consisted of about twenty people; maybe half of them students, the other part devotees and curious listeners. It was rather obvious from his reactions that he despised the fan club gathered to listen to the master's voice, but somehow he longed for them, too—very ambivalent, rather narcissistic and probably not unusual for prominent people.

After one of his lectures, I approached him and asked him if we could meet. He responded friendly but not very enthusiastically. I guess he was not really interested in a very young, naive physicist from Vienna. What cold I offer him despite boredom?

Alas, I had the impression that he was after the girls. I guess that if I would have been a pretty girl, I would have had very good chances of meeting him and have a chat or two. But again this is one of those counterfactuals I was speaking of before.

Feyerabend himself [33] and also Fischer [36] spoke about Feyerabend's sex life. Fischer recalls that Feyerabend told him he was beaten in his face (in German *"Ohrfeige"*) by his parents for putting his hands in his trouser pockets because they thought that he would masturbate [37]. For me it is quite remarkable that he was a womanizer on one side while on the other hand seemed not to have had a single coitus during his entire life span. I take this as an indication that there was deep dissatisfaction with this situation, a malady which was possibly not only caused by his war time injuries, but by psychic traumata which may have been deeply hidden and never showed up during his conscious phases in-between dreams and deep sleep. This may also be the ultimate reason for the kind of *"Unerfülltheit"* I observed, but of course I am wildly speculating here.

5.2 Science policy almost unimpressed

There was another issue which Feyerabend seemed to have taken quite light-heartedly; at least that was my impression: nobody took him seriously.

Paul Feyerabend had become almost a shooting star of philosophy of science, an icon of freedom and heresy to a generation coming of age in the late period of the twentieth century. Yet he never quite managed to obtain influence and convince scientific peers, governments and electorates to implement his recommendations regarding the selection of science funding and the implementation of science in general. In the lectures I attended during 1983, he strongly supported a system of lay judges for science assessment and financing; in closest analogy to the procedures established in the judicial system. I did never find so strong commitments to lay judges in his writings as I heard them in these lectures. I do not know and also did not dare to ask whether he was saddened by his lack of influence in practical terms. I would suspect that he had such a bad opinion about science mandarins and politicians that he expected not too much.

There may be evident and straightforward reasons why various funding agencies never seriously considered to adopt Feyerabend's proposals: Feyerabend's proposal would introduce an uncontrollable element in the distribution of money. For not only might quacks receive public funding; An even more disturbing consequence could be that, by a fairly independent selection of committee members and lay evaluators, powerful groups within the scientific community might loose their carefully crafted and delicately executed influence over the smooth flow of money towards them and their clients.

To some readers this may sound like a silly conspiracy theory. To these I respond that the matter is not obvious but quite serious and deserves careful attention of the general tax paying public, to which this article is not addressed. Let me just mention a large-scale study [17], in which 150 research projects of physics, chemistry and economic science were re-examined by the *National Science Foundation*. The results were devastating. This study showed how strongly the acceptance or refusal of a research project depends on the choice of the particular reviewer evaluating that proposal:

> An experiment in which 150 proposals submitted to the National Science Foundation were evaluated independently by a new set of reviewers indicates that getting a research grant depends to a significant extend on chance. ... the degree of disagreement within the population of eligible reviewers is such that whether or not a proposal is funded depends in a large proportion of cases upon which reviewers happen to be selected for it.

This finding is rather disturbing, as in many funding agencies, the "referents" in charge of selection of the peer reviewers are nominated in a rather unaccountable and certainly not very transparent manner.

As an attempt to implement Feyerabend's approach to methodology, I would like to suggest to distribute the funding for research projects in the following manner: First, make a very rough plausibility check to eliminate applications which are obviously fraudulent, inconsistent or otherwise impossible, unlawful or catastrophic. This could be done by lay judges. Then, in the first round of money distribution, choose 10% projects for funding totally at random. Choose the next 20% by a system of lay judges, as suggested by Feyerabend. Finally, distribute the remaining 70% via the conventional peer review process. After five years, publish the outcome of the projects funded by all three selection categories and adjust the relative magnitude of these categories accordingly. The outcomes might be quite amazing.

5.3 Bureaucratic dangers to science

There is one development which Feyerabend did not foresee: the growing detrimental dominance of the administrative bureaucracies over the scientists. This is administered through an ever increasing net of what appears to be checks and balances, of scientometric factors and numbers, of frequent evaluations and proposals and various certification and standardization procedures. This makes perfect sense for the administration, whose major task it is to distribute public money smoothly, accountably in terms of records, and free of risks. But this is not seldom opposed to science, and also opposed to the methodological openness Feyerabend had in mind. It also is quite frustrating for the researchers which are captives of this treadmill. The behemoth created by the Sixth Framework Program (FP6) and the establishment of the European Research Area (ERA) are such examples, but there are numerous others on local and institutional scales.

5.4 For creativity and abundance

Let me finish more conciliatory and stress the heritage of Paul Feyerabend. To me, of the many wise and weird things he said and wrote, two messages stand out. The first one is in the spirit of the Enlightenment and gets close to what also Kant had in mind: try on your own, let not others decide what you think; do not stop where other people, authorities and mandarins tell you to halt. And finally, in his last manuscript, Feyerabend calls upon us to reach out and conquer the abundance.

Acknowledgements and disclaimer

This paper grew out of a seminar held at the Institute of Philosophy of the University of Vienna jointly with Kurt Fischer and Anja Weiberg in the fall semester 2003/04. I am deeply indebted Professor Fischer for the discussions and insights he shared with me.

The views expressed here are my personal impressions and do not in any way reflect the official position of any of the institutions involved. Misinterpretations and errors are in the sole responsibility of the author.

References

[1] Angus Armitage, "The deviation of falling bodies", *Annals of Science* **5** (1941-47), 342–351.
[2] Alain Aspect, Philippe Grangier, and Gérard Roger, "Experimental tests of realistic local theories via Bell's theorem", *Physical Review Letters* **47** (1981), 460–463.
[3] _____, "Experimental realization of Einstein-Podolsky-Rosen-Bohm gedankenexperiment: A new violation of Bell's inequalities", *Physical Review Letters* **49** (1982), 1804–1807.
[4] _____, "Experimental test of Bell's inequalities using time- varying analyzers", *Physical Review Letters* **49** (1982), 1804–1807.
[5] John S. Bell, "How to teach special relativity", *Progress in Scientific Culture* **1** (1976), no. 2, Reprinted in [6].
[6] _____, *Speakable and unspeakable in quantum mechanics*, Cambridge University Press, Cambridge, 1987.

[7] _____, "George Francis FitzGerald", *Physics World* **5** (1992), no. 9, 31–35, Abridged version by Denis Weaire.
[8] Garrett Birkhoff and John von Neumann, "The logic of quantum mechanics", *Annals of Mathematics* **37** (1936), no. 4, 823–843.
[9] George Boole, "On the theory of probabilities", *Philosophical Transactions of the Royal Society of London* **152** (1862), 225–252.
[10] _____, *An investigation of the laws of thought*, Dover edition, New York, 1958.
[11] Percy W. Bridgman, *The logic of modern physics*, New York, 1927.
[12] _____, "A physicist's second reaction to Mengenlehre", *Scripta Mathematica* **2** (1934), 101–117, 224–234, Cf. R. Landauer [52].
[13] _____, *The nature of physical theory*, Princeton, 1936.
[14] _____, *Reflections of a physicist*, Philosophical Library, New York, 1950.
[15] _____, *The nature of some of our physical concepts*, Philosophical Library, New York, 1952.
[16] Maria Luisa Dalla Chiara and Roberto Giuntini, "Popper and the logic of quantum mechanics", Invited talk at the Karl Popper 2002 Centenary Congress, Vienna, July 5th, 2002.
[17] Stephen Cole, Jonathan R. Cole, and Gary A. Simon, "Chance and consensus in peer review", *Science* **214** (1981), 881–885.
[18] Paul A. M. Dirac, "Is there an aether?", *Nature* **168** (1951), 906–907.
[19] Freeman Dyson, "Unfashionable pursuits", *Mathematical Intelligencer* (1983), 47–54, Dyson gave related talks; e.g., at Yale, Minnesota and Bonn.
[20] Albert Einstein, „Zur Elektrodynamik bewegter Körper", *Annalen der Physik* **17** (1905), 891–921, English translation in the Appendix of [54].
[21] _____, *Äther und Relativitätstheorie. Rede gehalten am 5. Mai 1920 an der Reichs-Universität Leiden*, Springer, Berlin, 1920.
[22] Albert Einstein, Boris Podolsky, and Nathan Rosen, "Can quantum-mechanical description of physical reality be considered complete?", *Physical Review* **47** (1935), 777–780.
[23] Paul K. Feyerabend, *Against method*, New Left Books, London, 1974.
[24] _____, "How to defend society against science", *Radical Philosophy* **11** (1975), reprinted in [26].
[25] _____, "Hidden variables and the argument of Einstein, Podolsky and Rosen", *Problems of Empiricism* (Philosophical Papers, Vol 2), Cambridge University Press, Cambridge, 1981, pp. 298–342.

[26] _____, "How to defend society against science", *Scientific Revolutions* (Ian Hacking, ed.), Oxford University Press, Oxford, 1981, pp. 156–167.
[27] _____, "Niels Bohr's world view", *Problems of Empiricism (Philosophical Papers, Vol 2)*, Cambridge University Press, Cambridge, 1981, pp. 247–297.
[28] _____, "On the quantum theory of measurement", *Problems of Empiricism (Philosophical Papers, Vol 2)*, Cambridge University Press, Cambridge, 1981, pp. 207–218.
[29] _____, *Problems of empiricism. Philosophical papers, Volume 2*, Cambridge University Press, Cambridge, 1981.
[30] _____, "Professor Bohm's philosophy of nature", *Problems of Empiricism (Philosophical Papers, Vol 2)*, Cambridge University Press, Cambridge, 1981, pp. 219–235.
[31] _____, *Realism, rationalism and scientific method. Philosophical papers, Volume 1*, Cambridge University Press, Cambridge, 1981.
[32] _____, "Reichenbach's interpretation of quantum mechanics", *Problems of Empiricism (Philosophical Papers, Vol 2)*, Cambridge University Press, Cambridge, 1981, pp. 236–246.
[33] _____, *Killing time*, The University of Chicago Press, Chicago and London, 1995.
[34] Paul K. Feyerabend and John Preston, *Knowledge, science and relativism. Philosophical papers, Volume 3*, Cambridge University Press, Cambridge, 1999.
[35] Richard P. Feynman, *The character of physical law*, MIT Press, Cambridge, MA, 1965.
[36] Kurt Rudolf Fischer, „Paul Feyerabend: A personal reminiscence", *Aufsätze zur angloamerikanischen und österreichischen Philosophie* (Kurt Rudolf Fischer, ed.), Lang, Frankfurt am Main, Wien, 1999, pp. 77–89.
[37] _____, *Feyerabend's Weltanschauung*, Springer, Vienna, 2004.
[38] George Francis FitzGerald, *The Scientific Writings of the Late George Francis FitzGerald* (J. Larmor, ed.), Dublin, 1902.
[39] J. R. Greechie, "Orthomodular lattices admitting no states", *Journal of Combinatorial Theory* **10** (1971), 119–132.
[40] Daniel B. Greenberger and A. YaSin, "'Haunted' measurements in quantum theory", *Foundation of Physics* **19** (1989), no. 6, 679–704.
[41] Edwin H. Hall, "Do falling bodies move south?", *Phys. Rev.* (Series I) **17** (1903), 179–190.
[42] _____, "Do falling bodies move south?", *Phys. Rev.* (Series I) **17** (1903), 245–254.

[43] Heinrich Hertz, *Prinzipien der Mechanik*, Barth, Leipzig, 1894.
[44] Thomas J. Herzog, Paul G. Kwiat, Harald Weinfurter, and Anton Zeilinger, "Complementarity and the quantum eraser", *Physical Review Letters* **75** (1995), no. 17, 3034–3037.
[45] K. Hess and W. Philipp, "Exclusion of time in the theorem of Bell", *Europhysics Letters* **57** (2002), no. 6, 775–781.
[46] Clifford Alan Hooker, *The Logico-Algebraic Approach to Quantum Mechanics*. Volume I: *Historical Evolution*, Reidel, Dordrecht, 1975.
[47] Max Jammer, *The conceptual development of quantum mechanics*, McGraw-Hill Book Company, New York, 1966.
[48] _____, *The philosophy of quantum mechanics*, John Wiley & Sons, New York, 1974.
[49] _____, "John Steward Bell and the debate on the significance of his contributions to the foundations of quantum mechanics", *Bell's Theorem and the Foundations of Modern Physics* (A. van der Merwe, F. Selleri, and G. Tarozzi, eds.), World Scientific, Singapore, 1992, pp. 1–23.
[50] Simon Kochen and Ernst P. Specker, "The problem of hidden variables in quantum mechanics", *Journal of Mathematics and Mechanics* **17** (1967), no. 1, 59–87, Reprinted in [62].
[51] Imre Lakatos, *Philosophical papers. 1. the methodology of scientific research programmes*, Cambridge University Press, Cambridge, 1978.
[52] Rolf Landauer, "Advertisement for a paper I like", *On Limits* (John L. Casti and J. F. Traub, eds.), Santa Fe Institute Report 94-10-056, Santa Fe, NM, 1994, p. 39.
[53] A. A. Michelson and E. W. Morley, "On the relative motion of the earth and the luminiferous ether", *Amer. J. Sci.* **34** (1887), 333–345.
[54] Arthur I. Miller, *Albert Einstein's special theory of relativity*, Springer, New York, 1998.
[55] Eric Oberheim, "Bibliographie Paul Feyerabends", *Journal for General Philosophy of Science* **28** (1997), 211–234, An earlier german version of The Works of Paul Feyerabend.
[56] _____, "The works of Paul Feyerabend", *Knowledge, Science and Relativism. Philosophical Papers*, Volume 3 (John Preston, ed.), Cambridge University Press, Cambridge, 1999, An improved and updated bibliography of the works of Paul Feyerabend., pp. 227–251.
[57] Asher Peres, "Karl Popper and the Copenhagen interpretation", *Stud. History and Philos. of Modern Physics* **33** (2002), 23–34.

[58] Erwin Schrödinger, „Die gegenwärtige Situation in der Quantenmechanik", *Naturwissenschaften* **23** (1935), 807–812, 823–828, 844–849, English translation in [79] and [80]; http://www.emr.hibu.no/lars/eng/cat/.
[59] R. S. Shankland, S. W. McCuskey, F. C. Leone, and G. Kuerti, "New analysis of the interferometer observations of Dayton C. Miller", *Rev. Mod. Phys.* **27** (1955), 167–178.
[60] Alan Sokal, http://www.physics.nyu.edu/faculty/sokal/.
[61] Ernst Specker, "Die Logik nicht gleichzeitig entscheidbarer Aussagen", *Dialectica* **14** (1960), 175–182, Reprinted in [62]; English translation: "The logic of propositions which are not simultaneously decidable", reprinted in [46].
[62] ____, *Selecta*, Birkhäuser Verlag, Basel, 1990.
[63] Walter Terence Stace, "The refutation of realism", *Readings in philosophical analysis* (Herbert Feigl and Wilfrid Sellars, eds.), Appleton–Century–Crofts, New York, 1949.
[64] Susan G. Sterrett, "Sounds like light: Einstein's special theory of relativity and Mach's work in acoustics and aerodynamics", *Studies In History and Philosophy of Science* Part B: *Studies In History and Philosophy of Modern Physics* **29** (1998), 1–35.
[65] Karl Svozil, "On the setting of scales for space and time in arbitrary quantized media", Lawrence Berkeley Laboratory preprint **LBL-16097** (1983), http://heplibw3.slac.stanford.edu/spires/find/hep? key=1089510 a pdf scan is at URL http://ccdb3fs.kek.jp/cgi-bin/img/allpdf? 198309187.
[66] ____, "Connections between deviations from Lorentz transformation and relativistic energy-momentum relation", *Europhysics Letters* **2** (1986), 83–85, excerpts from [65].
[67] ____, *Randomness & undecidability in physics*, World Scientific, Singapore, 1993.
[68] ____, "A constructivist manifesto for the physical sciences", *The Foundational Debate, Complexity and Constructivity in Mathematics and Physics* (Dordrecht, Boston, London) (Werner DePauli Schimanovich, Eckehart Köhler, and Friedrich Stadler, eds.), Kluwer, 1995, Cf. [69], pp. 65–88.
[69] ____, "How real are virtual realities, how virtual is reality? The constructive re-interpretation of physical undecidability", *Complexity* **1** (1996), no. 4, 43–54.
[70] ____, *Quantum logic*, Springer, Singapore, 1998.
[71] ____, "Relativizing relativity", *Foundations of Physics* **30** (2000), no. 7, 1001–1016, e-print arXiv:quant-ph/0001064.
[72] ____, *The dangerous misconceptions of Sir Karl Raimund Popper*, 2002.

[73] _____, Logical equivalence between generalized urn models and finite automata, 2002.
[74] _____, What could be more practical than a good interpretation?, 2002.
[75] _____, On counterfactuals and contextuality, 2004.
[76] _____, "Quantum information via state partitions and the context translation principle", *Journal of Modern Optics* **51** (2004), 811–819.
[77] Karl Svozil and Josef Tkadlec, "Greechie diagrams, nonexistence of measures in quantum logics and Kochen–Specker type constructions", *Journal of Mathematical Physics* **37** (1996), no. 11, 5380–5401.
[78] T. Theocharis and M. Psimoloulos, "Where science has gone wrong", *Nature* **329** (1987), 595–598.
[79] J. D. Trimmer, "The present situation in quantum mechanics: a translation of Schrödinger's 'cat paradox'", *Proc. Am. Phil. Soc.* **124** (1980), 323–338, Reprinted in [80].
[80] John Archibald Wheeler and Wojciech Hubert Zurek, *Quantum theory and measurement*, Princeton University Press, Princeton, 1983.
[81] Ron Wright, "The state of the pentagon. A nonclassical example", *Mathematical Foundations of Quantum Theory* (A. R. Marlow, ed.), Academic Press, New York, 1978, pp. 255–274.
[82] _____, "Generalized urn models", *Foundations of Physics* **20** (1990), 881–903.

Notes

1. email: svozil@tuwien.ac.at, homepage: http://tph.tuwien.ac.at/~svozil
2. Hlavka reports that Hans Thirring and Hahn conducted parapsychological experiments in an apartment at the *Ringstrasse*.
3. „Wir machen uns innere Scheinbilder oder Symbole der äußeren Gegenstände, und zwar machen wir sie von solcher Art, dass die denknotwendigen Folgen der Bilder stets wieder die Bilder seien von den notwendigen Folgen der abgebildeten Gegenstände. ... In der That wissen wir auch nicht, ob unsere Vorstellungen von den Dingen mit jenen in irgend etwas anderem übereinstimmen, als allein in eben jener *einen* fundamentalen Beziehung."

JULIET FLOYD

HOMAGE TO VIENNA: FEYERABEND ON WITTGENSTEIN
(AND AUSTIN AND QUINE)[1]

In this essay I shall be examining, not so much the question of whether Feyerabend's epistemological anarchism and his arguments on behalf of theory-proliferation are coherent or attractive views, but, rather, how his articulation of these positions fits into the evolution of analytic philosophy during the post-war period, and in particular, into Feyerabend's own sometimes quixotic assimilation of Wittgenstein. Thus, although Wittgenstein's impact (and lack of impact) upon Feyerabend provide the occasion and backdrop for this paper, I shall be broadening the context by contrasting Feyerabend's attitudes toward Wittgenstein, Austin and Quine with my own. My aim is to set Feyerabend's remarks about Wittgenstein and his embrace of anti-method into a richer philosophical and historical background context than is usually done. Too often Feyerabend's philosophy has been assumed to have been primarily directed at Popper's critical rationalism, and the impact of Wittgenstein upon him to have either been negligible or very general, turning on a picture of the conceptually-saturated nature of experience, the "disunity" of science, or an "end of philosophy" thesis. By widening our perspective on these assumptions to take into consideration further details of the historical context to which they apply, we can, I shall argue, better understand Feyerabend's philosophical significance. But, more generally, this essay will set forth, by way of example, some of my own proposals about why understanding the history of analytic philosophy historically is interesting, important, and relevant to philosophy's development today. Since my own perspective on the development of analytic philosophy contrasts rather sharply with recent narratives that have been proposed by Dummett, Hacker, Rorty and Soames[2], I shall be using my discussion of the limitations of Feyerabend's perspective to make a number of positive suggestions about how we might approach a re-reading of certain canonical figures of the recent philosophical past, and the figures of Austin, Quine, and Wittgenstein in particular. One of the instructive things about Feyerabend is that his partial appropriations of (and his blindnesses about) Austin, Quine and Wittgenstein, precisely because they are not atypical, provide us with further insight, as it were by interpolation or subtraction, into what makes each of these other three so important to the history of twentieth century analytic philosophy.

The evolution of analytic philosophy as a whole is obviously a large and contested terrain, partly because it is insufficiently understood in historical terms by those who write within and about it: analytic philosophers have always tried to draw a sharp boundary between the history of philosophy and what they do. Feyerabend was an analytic philosopher who had a richer conception of the historical evolution of philosophy than most. My method here is to use him as a stepping-stone into a juggling act in which I shall try to keep a number of different philosophers and themes in the air all at once. I shall insert at points my own (sometimes revisionary) remarks about Feyerabend, Austin, Quine and Wittgenstein, placing these alongside some of their remarks about one another. Different correctives will be offered at different points by different voices, all in an effort to bring what I regard as some of most important philosophical problem context(s) and themes to the fore. I shall not be defending my interpretations so much as arranging them, thereby urging upon my readers the need for a larger narrative about the history of analytic philosophy, a narrative that does not focus on one or two ideas, philosophical movements, or figures in isolation, but looks toward the painting of a wider, more complicated canvas involving many different thinkers, traditions and themes. Limitations of time and space require me at times to make claims about each philosopher that I am well aware will strike some readers as revisionary, and ought, ideally, to be argued for in detail. My idea is, however, not simply to shock and provoke, but instead, to use each philosopher in the foursome on behalf of the other, where possible, in a productive dialogue, drawing out from them some of what I take to be most lasting in their philosophies. For this reason, I adopt no fixed linear ordering among the names in the group, but place one name before another for emphasis, as needed in a particular context, as a way of acknowledging my own application of one philosopher's perspective to another. Although I shall say least about Austin – a philosopher who, among the four I discuss, may appear to require most defense as a relevant figure – this does not reflect my view of his relative importance, so much as my failure to have as yet found a way to put his insights satisfactorily to use in my own thinking. As the attentive reader will see, I hope to be able to do more of this in the future. Here I aim to convey primarily that I take Austin to be the most vastly underrated philosopher in the group, at least by the members of the group, and I sketch some suggestions about why that might be so.

My comparative approach will allow me to explore the question of what it is or might be for one philosopher to be "deeper" or "richer" or more historically important or more instructive than another, and also to enact and explore several important ambiguities facing the whole idea of anti-method in philosophy, something I take to be, in the hands of each of the four philosophers I shall discuss, something more than a jejune relativism, pluralism, epistemological opportunism, or skepticism. Under at least one constructive conception of anti-method, I applaud it, both in philosophy and in science; it seems to me what remains of empiricism, when it has been purged of its dogmas, and an important conception of how philosophy may hope to survive in an age of increasing fragmentation and specialization. It seems to me therefore that anti-method has sometimes been too narrowly conceived (as a localized rejection of the hypothetico-deductive method, and/or Popper's emendation of it, in philosophy of science) and sometimes too broadly (as an embrace of irrationalism, "post-modernism", "anti-theory", or a general "end of philosophy thesis"). For it has also, as I shall be arguing, sometimes been effectively embraced and furthered where it has been neither explicitly enunciated nor seen to have been at work – even by Feyerabend himself, its most accessible and enthusiastic promoter.

1. Feyerabend on Wittgenstein: Background to Some Blindnesses

If we are to believe his testimony, Feyerabend seems to have admired Wittgenstein above all other philosophers of the twentieth century.[3] But we shall have to probe what profit he derived from this admiration, and what form(s) it took. Feyerabend was liable to distance himself, in the style of Popper and Gellner, both from Wittgenstein and from Austin, especially where their philosophies were connected with something he called "ordinary language philosophy". At the same time he was also willing to except Austin and Wittgenstein as individuals from this condemnation. He was also liable to invoke the influence of Wittgenstein upon him when he wished to avoid having a position, theory, or "ism" pinned upon him, or when he wished not to be classified as a member of Popper's circle.[4]

At a very general level, there are obvious ways in which Wittgenstein and Feyerabend are kindred spirits. They each took for granted that there is a certain progressive scientific spirit informing Western scientific culture, what Wittgenstein called (in 1930) "the vast stream of European and American civilization in which all of us stand" (N.B.: Wittgenstein is not exempting himself from this dwelling, whatever longing he might also express, at times, for an escape from it).[5] And they each stressed that dogmatism, prejudice, and conceptual unclarity appear within scientific practice just as regularly as they do everywhere else in human life. A major philosophical challenge for human beings in the modern world is to learn how to treat science's increasingly complex claims with appropriate respect and understanding, but without ever converting them, or any philosophy that critiques them, into a secularized religion or *Weltanschauung*, even (and especially) unwittingly, as Wittgenstein and Feyerabend each felt may have happened in the case of pragmatism.[6] It is therefore part of the task of philosophy, as Wittgenstein and Feyerabend conceive it, to investigate and critique certain symbolic, dogmatic, and/or empty pronouncements that arise about and within science, but without preaching about "truth" from on high, dreaming of a purified scientific method or language, or attempting to speak in general terms about technological thinking or science as such.[7] Feyerabend's idea that good philosophical writing can and ought to be opinionated, impressionistic, informed yet anti-scholastic exemplifies this conception.

Their commonality of spirit is also reflected in the fact that Wittgenstein and Feyerabend each emphasized multifariousness and complexity – on different grounds, in different circumstances, and to different effect – by highlighting the fact that if we look and see, there are to be found a many-colored variety of techniques, activities, productions and speculations within scientific and philosophical practice, rather than one overarching method or doctrine or theory called "science".[8] No philosopher stands upon a mountain with a clear view of the whole, above and beyond knowledge viewed through the eyes of a participating, implicated, partial, historically situated, fallible, yet not wholly ignorant, knower.

The ideas attending and motivating this emphasis on many-coloredness – an appropriate emphasis for an age of increasing fragmentation and specialization in all branches of human inquiry – are many,[9] and when we begin to articulate them, significant commonalities with the philosophies of Austin and Quine begin to become apparent where they may not have been before. I shall put these ideas in my own voice, as I feel that each philosopher in our foursome brings home their force, however differently, simply by taking them as starting points – something not all philosophers do.

It is intrinsic to the pursuit of truth that we pursue not only truth, that we pursue truth always in one or more contexts, and that as we pursue it we face discomfort about our aims, motives, and effects, as well as conceptual groping and unclarity about our notions and concepts. "Knowledge" should be pondered as a human, not a Godlike, cosmologically-indexed or acontextually analyzeable phenomenon.[10] What Vienna positivism failed sufficiently to stress in this regard is that there is (so far, at least) no general theory of meaning and no general philosophical method or position that can innoculate one against all intellectual follies ahead of time – not, anyway, if one is a thinking, reasoning human being who does progress in understanding. Nonsense, conceptual confusion and dogmatism are therefore not, as phenomena, just Wittgensteinian or positivistic philosophy's constructions, they are intrinsic to the human pursuit of knowledge and understanding, and as such they form a diverse family. The positivistic dream that you might avoid or tame meaninglessness once and for all by means of a theoretical or methodological recipe, or that confusion and dogma are unimportant obsessions for Wittgensteinian philosophers alone (byways not liable to be encountered within properly-pursued scientific method) are just grave intellectual mistakes.

This is not to dismiss truth or theory-construction as unimportant. Truth is obviously essential, and rightly thirsted for, it is precious, sometimes beautiful, often useful and easily obscured. It is not all that easy to obtain in certain cases, and one had better not run afoul of what is known of it, or might more or less easily be known of it, when one philosophizes. Moreover, there are philosophers who have contempt for truth and for science (by which I mean the English sense of the word "science", and not *Wissenschaft*): they set up *a priori* reasons why what they say cannot be discussed or scrutinized in an open, intersubjectively shareable intellectual light, and neither Feyerabend nor Wittgenstein nor Austin nor Quine, as I understand them, had much use for this. If what I believe belongs to me or my culture alone, then that should be a useful, hard-won discovery, neither an *a priori* starting point nor an ending point: and it may best be an insight gained through conversation with others. Philosophy should aim to be cooperative where it can, and understand where and why and on what occasions it cannot be.

Nevertheless at its best philosophy, as it appeared both to Feyerabend and to Wittgenstein, as well as to Austin and to Quine, has primarily to do with inculcating understanding and judgment, and that is something different from the direct search for truth, whether spiritual, metaphysical, or empirical.[11] There is a lot of truth out there for the having. But one point of pursuing philosophy is to discover which of it you and/or others ought to seek and embrace, which of it you may hope to alter, which you should express (and how, and on what occasions) and which you may safely ignore. Differently put, philosophy has more to do with understanding which truths are the best and most important ones to enunciate in a given context, which arguments or analogies are worth getting involved in, and which are likely to distract human beings from what is most important in a given context. One aim is to achieve some degree of self-consciousness about the effects of one's words and arguments upon others and upon oneself.

It is important to emphasize, contrary to what is all too frequently supposed, that neither Wittgenstein nor Feyerabend embraced an "end of philosophy" thesis.[12] That is to say, neither one held, as I understand them, that there is not or can no longer be such a thing as progress in philosophy, or that philosophy is insulated *a priori* from scientific practice, or that scientific knowledge is intrinsically bad and misleading, primarily deserving philosophical prosecution before the *a priori* bar of grammar and/or the *a posteriori* bars of psychology, sociology and history.[13] Feyerabend and Wittgenstein were constructive thinkers, not only debunkers. As I understand them, they were neither wholly anti-scientific, nor primarily anti-scientistic in their aims. Their philosophies just are more complex. Indeed, it seems to me that each of them sought and found forms of progress in places where most philosophers, scientists and mathematicians would not dream of looking for them.

Wittgenstein's and Feyerabend's fruitfulness turned, however, almost wholly on their abilities as model-, narrative- and analogy-builders, rather than on the re-classification of problems and conceptions according to more or less systematic sorting of verbal and grammatical structures. Quine and Austin contributed powerful analogies and metaphors as well, but they did not rest quite so much of their work on these. As Wittgenstein said to Waismann in a conversation about philosophical method in the early 1930s, "to make the unclarities and vagueness of our words perspicuous, one might exhaust oneself in devising pictures and similes".[14] Here we touch on the question of the kinds of stylistic similarities and contrasts that may be detected between Wittgenstein and Feyerabend, since it is obvious that the form in which each thinker wrote is internal to his philosophical power and ambition, and that Feyerabend's writing more or less explicitly avails itself of recurrent allusions to Wittgenstein's writings. I shall be touching upon the aesthetic preferences of Feyerabend and of Wittgenstein below, for these are indicators, as I see it, of some of the philosophical affinities and contrasts between them.

Feyerabend came, toward the end of his life, vehemently to denounce the idea that his philosophy was influenced by Popper's, precisely by insisting on the importance of Wittgenstein to his thought. The following remarks from an interview, published in 1991, give a flavor of this:

B: I didn't know Popper had a philosophy.
A: You cannot be serious. You were his student ...
B: I listened to some of his lectures ...
A: And became his pupil ...
B: I know this is what Popperians say ...
A: You translated Popper's *Open Society* ...
B: I needed the money ...
A: You mentioned Popper in footnotes, and quite frequently ...
B: Because he, and his pupils begged me to do so and I am kindhearted. Little did I know that one fine day such friendly gestures would give rise to serious dissertations about 'influences'.
A: But you were a 'Popperian' – all your arguments were in the Popperian style.
B: This is where you are mistaken. It is quite true that some of my discussions with Popper are reflected in my early writings – but so are my discussions with Anscombe, Wittgenstein, Hollitscher, Bohr, and even my reading of Dadaism, Expressionism, Nazi authorities has left a trace here and there. You see, when I come across some unusual ideas I try them out. And my way of trying them out is to push them to the extreme. There is not a single idea, however absurd and repulsive, that has not a sensible aspect and there is not a single view, however plausible and humanitarian, that does not encourage and then conceal our stupidity and our criminal tendencies. There is much Wittgenstein in all my papers – but Wittgensteinians neither seek nor are in need of great numbers of followers and so they do not claim me as one of their own. Besides they understand that while I regard Wittgenstein as one of the great philosophers of the twentieth century ...
A: Greater than Popper?
B: Popper is not a philosopher, he is a pedant – that is why the Germans love him so. At any rate – the Wittgensteinians realize that my admiration for Wittgenstein does not yet make me a Wittgensteinian. But this is all beside the point ...[15]

There are two thoughts expressed here that Feyerabend connects specifically with Wittgenstein. First, there is the idea that there is a difference between Wittgenstein and Wittgensteinians, between one individual philosopher and a philosophical movement claiming that philosopher as a master. That we should be, in Quine's phrase, "extensionalists" in the history of philosophy, reading philosophers one by one, as charitably and constructively as we can, rather than as types sorted by previously clear concepts or subject matters, is an idea whose importance I have appreciated since I first read *Against Method*.[16] After reading Feyerabend, one could imagine viewing the history of philosophy as a kind of intrinsically interesting dramatic dialogue among more or less well-informed individuals – subject to a variety of intellectual and existential pressures and limitations, as individuals always are – about the human effort to articulate a picture of the world and our place in it.

Second, and less like Wittgenstein, is Feyerabend's thought that "there is not a single idea, however absurd and repulsive, that has not a sensible aspect, and there is not a single view, however plausible and humanitarian, that does not encourage and then conceal our stupidity and our criminal tendencies". Feyerabend's well-known principle of epistemological anarchism, that "anything goes", an adaptation of Millean concerns about the tyranny of majority opinion, is related to an (I think UnMillean) idea of the incipient bestiality of all of our thoughts and views. This is part of what is enunciated in *Against Method*.[17] Kurt Fischer has suggested that "anything goes" may be viewed as an extension of a kind of principle of tolerance across the board, to all approaches within and outside of science and philosophy. It may also be taken to function primarily negatively, as a dialectical tool designed to undercut uncritical rationalism and philosophical absolutism.[18] Fischer also connects Feyerabend's anarchistic principle with the Nestroy quotation that forms the motto to Wittgenstein's *Philosophical Investigations*, "progress has this in it, that it always seems greater than it is". His suggestion here is that the heart of Feyerabend's anti-method is an insistence on the piecemeal nature of human progress, on its occurrence by means of small steps, rather than global insights. And this certainly squares with Feyerabend's repeated emphasis on the importance of what he called "practice" over "theory", instantiated by his experiences with non-Western medicine and cuisine.[19]

As I shall suggest below, however, Feyerabend's attitude toward Wittgenstein's motto from Nestroy is more complicated, and his view of Wittgenstein more ambivalent. For one thing, distinctions between "practice" and "theory" and piecemeal vs. global progress are alien to

Wittgenstein's philosophy, as, importantly, they are not to Popper's. It must be said that Popper shaped Feyerabend's writings as much as did Wittgenstein, despite the avowed lack of influence suggested in the above quotation. Feyerabend's denunciations of the idea of a sharp distinction between theoretical terms and observation terms, his antagonism toward what he called "Popper's Church"[20], his insistence on the importance of theory-proliferation, of pluralism, and of anti-method, his conception of the continuity between philosophy and the arts, may each be seen to apply more or less directly to Popper. Of course, while Feyerabend's anti-method stance was shaped by Popper, it is also, in *Against Method*, directed at Lakatos, to whom it was written as a "letter", as Feyerabend explains in his prefatory remarks. (Lakatos was to have written the other half of the book as a reply, but died before he could do so.[21]) In more ways than one, *Against Method* is a half-formed composition, missing one hand. That this did not prevent it from becoming a cult classic is a testimony, I suppose, to its author's dramatic gifts as a writer of theatrical history of science – a point to which I shall return in the next section of the paper.

Having sampled a few of Feyerabend's later remarks about Wittgenstein, it behooves us to probe the extent to which this is or is not a re-writing of Feyerabend's own history. It also invites us to explore the extent to which Feyerabend's own thoughts reflect a larger historical and philosophical matrix of ideas, ideas that have their history, their value and their future, as well as their limitations.[22]

Feyerabend's most important localized contributions lay in the philosophy of physics. This was less because he had anything of unique and lasting importance to say about the Copenhagen interpretation of quantum mechanics than because he so passionately insisted on the importance of philosophers having something intelligible to say about it. Physics required philosophy, for Feyerabend, and not merely the other way around. He refused to accept the pronouncements of physicists who would insist that there are no important philosophical questions about quantum mechanics, or none that might not be solved in a month. The "prose" of the physicists implicates all of us, on Feyerabend's view. Like Wittgenstein, Feyerabend refused to be cowed by the fear of appearing ridiculous in probing scientist's characterizations of their own views and theories.

Of course, the risks, lessons, ambiguities and liabilities of this view are many, and I cannot hope to characterize them all here. So I shall have to be brief, given the framework of the present essay. As I see it, Feyerabend's strengths and limitations as a philosopher stemmed from the way he picked up certain pieces of the Viennese philosophical tradition after the Second World War and rearranged them in an original, but partial, way. From Mach Feyerabend learned to conceive physics as an historical tradition continuous with philosophy and the nature of sensation as a principal philosophical field.[23] By observing Ehrenhaft's experimental ingenuity he learned the importance of openness to practice and scepticism about current theory in the pursuit of experimental science, as well as respect for the complexity of the role of observation, especially of unintended empirical effects (*Dreckeffekten*) in the generation of empirical evidence.[24] From the Viennese positivist tradition he took the idea of philosophy of science as a core area of philosophy, both in the philosophical problems that are to be understood as central and in dictating philosophy's cultural importance within modern industrialized society. At the same time, from his aesthetic experiences in Vienna he imbibed a deep appreciation of the Austrian literary, musical and operatic traditions, taking drama, and the history of art more generally, to be part and parcel of philosophy, an aspect of our natural history, so to speak.[25] Anyone who has observed how carefully paired are the collections standing across from one another in Vienna's museums of Natural History and Art History – designed, as they are, to comment upon one another's displays of the natural – will understand how it could have been as obvious to Feyerabend as it was to Wittgenstein to conceive philosophy and science as implicated in the history of artistic stylization, and to find unholy the empiricist's myth of the uninterpreted given. What would remain of empiricism in Feyerabend (as in Wittgenstein) would be a sense of respect for attention to particular cases, for the stubbornness and the complexity of observation, rather than any particular account of the role of perception in thought.

The thread through Feyerabend's whole life's work is his attack on a sharp distinction between theory and observation, an insistence on the conceptually- and psychologically-saturated character of experience. Beside this is, however, another thread that he constantly misread and underestimated: the myths about thought, concepts, meaning and language into which we are liable unwittingly to fall in mounting this attack. We may grant that meaning and conceptualization matter to discussions of the evidentiary role of perception, but still want to know how they do. Are our notions of "language", "theory", "concept", "meaning" and "experience" clear in advance of an investigation? Feyerabend sometimes almost bordered on grasping the significance of such questions. When he told stories, he avoided the trouble. He also demonstrated a laudable refusal to define philosophy, either by method or subject matter. But, in the end, his preoccupation with Popper's form of rationalism took up more of Feyerabend's time than was good for him, skewing and truncating his engagement with Wittgenstein.

Feyerabend and Kuhn are usually viewed as the arch critics of the hypothetico-deductive model of scientific progress, critics whose refutation turned on the idea of the theory-ladenness of observation. But Kuhn appropriated more directly or explicitly than did Feyerabend those parts of Wittgenstein's philosophy that would last: the notion of *paradigm* is one of them. Though Feyerabend did eventually come to understand that Kuhn's notion of *paradigm*, like Wittgenstein's, is a family-resemblance notion whose applications vary tremendously from case to case,[26] in contrast to Kuhn he missed some of the most significant aspects of Wittgenstein's attack on the category of *a priori* knowledge. He also spent far too long defending "isms" in philosophy. These were legacies of the Popperian context. Late in his life he wonders why it took him so long to see through these limitations of his work.[27]

One reason, I suggest, is that, unlike Kuhn, Feyerabend had imbibed from the Vienna positivists the tendency to speak uncritically about meaning – even and perhaps especially while railing against theories of meaning. By the time Feyerabend became a student, the new logic was already a well-established philosophical instrument and institutionalized discipline, and ideas about the notion of meaning deriving from the Vienna Circle had been absorbed into the heart of academic philosophy in the Anglo-American and the Austrian worlds. Appeals to "rules of language" and "questions of meaning" to solve (or dissolve) philosophical problems were recognized professional philosophical techniques in the circles in which Feyerabend moved. It was often assumed that language is *mere* language, a structure and a subject matter given *a priori*, and hence available for analysis that would be free of any specific empirical or ontological commitment. Analysis itself was taken reductively, to be based on the idea that language alone constitutes the primary subject matter for philosophy.

Feyerabend remained to the end a Popperian: he distrusted talk about language, particularly talk about language as a conceptual scheme within which we are locked and cannot escape without rupture: the so-called "myth of the framework".[28] Perhaps more importantly, Feyerabend distrusted the study of logic as central to our philosophizing about theories and language. But then he just gave up on thinking about the notion of meaning and about philosophy of logic generally in any systematic or critical way. Thus, unfortunately, he missed the heart of Wittgenstein's philosophy. And he remained out of step with the interests and accomplishments of some of the most significant philosophers who were to follow him in the next generation. Feyerabend recognized that Wittgenstein and Austin were philosophers centrally concerned with investigating both conceptions of meaning and of language, but these were topics that Feyerabend himself thought of as uninteresting. Thus, like Popper and Gellner, he wrongly associated Wittgenstein and Austin with an unhelpful, dogmatic relativism and with reactionary quietism.

Feyerabend was stubbornly unwilling to contemplate the idea that concepts like "real", "reality", "theory", "language" and "experience", far from remaining fixed across philosophical debates and divides, tend more than one might at first think to be warped, rather than simply applied, in the to and fro of philosophical argument. He never did take seriously the notion – more or less evident on the surface of Wittgenstein's *Philosophical Investigations* – that to gain an understanding of the complexity of our notions of language, name, proposition and so on we require a plurality of kinds of conceptual investigation, investigations that are difficult partly because of the surprising variety of what we do say, why, and when. Feyerabend never respected sufficiently the idea that this variety and complexity are tremendously difficult to see rightly, and that this fact is itself of central philosophical importance.[29]

Feyerabend regarded Wittgenstein as too ahistorical, too preoccupied with criticizing (rather than defending) myth, and too uncritical of "ordinary" or "common sensical" belief. He felt that Wittgenstein did not sufficiently appreciate that science progresses sometimes through the creation of myths.[30] But this was because Feyerabend did not himself sufficiently focus on myths of and about our language.

It is true that the later Wittgenstein often investigates imagined language-games and examples, simplified portions of language use, rather than actual historical events and persons. Yet it is also true that he nearly as often alludes to, hence presupposes, actual historical cases – as in his remarks on the foundations of mathematics, which are evidently preoccupied with Dedekind, Frege, Russell, himself, Hilbert, Gödel and Hardy, among others.[31] The irony is that it was Feyerabend, and not Wittgenstein, who misread the historical-philosophical context in which he lived, and precisely by vastly underestimating and miscasting the importance to philosophy, not only of the rise of modern mathematical logic, but of philosophy's distinctively new preoccupation, in the wake of this rise, with the complexity, variety and subtlety of forms of human speech and communication. It is as if, during the time of Galileo, one had dismissed the doing of astronomy as an idle pastime, refused to look through a telescope, and supposed that the history of painting had reached its aesthetic endpoint in a painter like Giotto (for more on Feyerabend on Giotto, see section 2. below). The new logic was a lens for newly refracted observations about language, but Feyerabend chafed at it like a philosophical straitjacket. This was an Austrian thing. Popper fell into it too. Yet, even by the time of the *Anschluss* the driving core of Vienna's high tradition of scientific philosophy had largely emigrated, and in its greater part toward the Anglo-American world, a world in which the figures of Wittgenstein, Austin and Quine (one might also mention here Gödel and Feigl) were moving toward a new-found appreciation of the complexity of human powers of expression and of the specific effects upon them of philosophical preconceptions about language and meaning. Feyerabend, like Popper, missed this boat. They each missed the multifariousness of the ways in which modern formal logic would serve as a new lens for philosophy, illuminating and distorting its questions in new kinds of ways.

For Frege and Russell did not just create modern mathematical logic, a new tradition (or set of traditions) in twentieth century mathematics, and a philosophically idle, purely formal, pastime. They also – as Wittgenstein, to his everlasting intellectual credit, saw early on – created the possibility, perhaps even the necessity, of new forms of philosophical confusion, deeper and more difficult to expose than the excesses of traditional metaphysics. It was Kant who had written that there is only so much science in a subject as there is mathematics, and on this ground the good Immanuel doubted that psychology would ever become a science.[32] A century later, Frege and Russell appeared to have mathematized logic, the most general laws of thought and reality. This appeared (in the words of the young Wittgenstein) to have revolutionized logic in something like the way in which modern chemistry revolutionized alchemy.[33] And yet these mathematical structures, this now fully genuine science (according to Kant's criterion) masqueraded as *a priori* metaphysics, rather than (what had been usual in the tradition before and after Kant) the other way around. Frege's and Russell's philosophical pronouncements – about the universal applicability of logic, about the way in which psychology is to be distinguished from logic, about truth, meaning and objectivity – were therefore much more weighty and apparently sophisticated, yet much more conceptually confusing and complicated to grasp, than prior metaphysics. Open up Kant or Hegel, and it is obvious that it is not mathematically articulated science. Open up Frege's *Grundgesetze* or Whitehead and Russell's *Principia Mathematica*, and it is obvious that at least some of it is. But it is then terribly unobvious where the prose begins, and the poetry, or genuine mathematics, ends – where, for example, the logic is a helpful organon for thinking about the nature of language or thought, and where, misinterpreted, it presents us with a mythological image of language, meaning and mind.[34]

That it is an important, difficult, and new form of intellectual project to meditate on the special difficulties and newfangled philosophical challenges here is something that Feyerabend was surprisingly unable to appreciate. He was also surprisingly unable to appreciate the force of the questions raised, both by Moore and by Austin (however differently) about the unobviousness and interest of that which we take to be obvious and boring (for example, our usual ways of expressing certainty and uncertainty in everyday speech) and about the special philosophical difficulties facing those who would try to articulate themselves plainly and clearly about the plain and the clear, precisely so as to demonstrate the unclarity of the apparently clear. What is hardest to see, as Wittgenstein remarked, is that which we take to be most obvious. And there are an enormous number of ideas about language that we tend to take to be clear and obvious at the start.

Feyerabend's failure to grasp the significance of such matters meant that his ways of establishing the theoretically-saturated nature of observation did not engage, as the philosophies of Austin, Quine, Wittgenstein did, with some of the most fundamental problems of his philosophical time. At the very least, his historical myopia blunted the force of his work. As I see it, properly viewed, analytic philosophy was a largely anti-Cartesian, anti-Idealist, anti-mentalist tradition, and, at least at its inception – in the work of Frege, Russell and Moore – it was not all that interested in language or method *per se*, but, rather, in propositions and truth conceived as ultimate ontological denizens of reality, and absolutely direct, unaffected and explicit assertion and acknowledgement of them as the norm.[35] At its origins analytic philosophy was not preoccupied with eliminating metaphysics, nor with scholastic hairsplitting of "mere" language, nor even with developing a "theory of meaning", much less with Cartesian epistemological certainty and the unadulterated, de-conceptualized representation of world by the mind – even if each of these narratives were to find a place within the tradition after the Vienna positivists. At root analytic philosophy was and remains obsessed with how to demarcate and systematically represent the contents of knowledge as *contents*, that is, as propositions, statements, sentences or assertions articulable and applicable quite apart from the activity of any particular person or group of persons, or any contingencies given by particular social, historical or psychological factors.

Thus I differ in emphasis and outlook with some of the most well-known large-scale narratives so far proposed by able historians of early analytic philosophy.[36] The analytic tradition's real trouble – a trouble analytic philosophy still struggles with today – is that new and often more obscure forms of essentialism, Cartesianism, and mentalism are likely to be produced in the very effort to extricate oneself from them by way of theories of objectivity that retain a problematic expressive ideal of clarity, an ideal holding that norms of speech and meaning are to be conceived acontextually, and expressions of thought as always articulable in declarative statements, true or false. This is something that Wittgenstein, Quine and Austin each deeply appreciated, each in his own way. Feyerabend also did, but without respecting certain complexities facing the philosophical effort to treat the notions of meaning and truth contextually, as historically realized notions. His own prejudices against the philosophical study of logic and language held him back. He seems to have too often presupposed that consideration of language could only enter into philosophy in a reductionist way, according to a positivistic view that all philosophical problems are only (and merely) problems of language and/or meaning. He therefore remained closed to the importance of such considerations by his own prejudices against the philosophical study of logic and language. While this did not detract everything worthwhile from his vivid historical narratives, it did, to repeat, lessen the force of their philosophical relevance within the context in which he was writing.

As Fischer reports, although Feigl and Tarski were instrumental in helping him get appointed at Berkeley,

Paul himself despised mathematical logic. He was interested in the accumulation of knowledge and he was not interested in its clarification – anyway, not by means of axiomatization – nor was he interested in its careful and tidy presentation. *That* one could easily surmise and feel, and one heard him say that on many occasions. He considered logic some sort of pastime, some kind of intellectual loitering, so to speak, with no results. He advised me not to study logic but mathematics; specifically that part of mathematics that is useful for an understanding of physics.[37]

Fischer continues:

> I admired the fine Anglo-Saxon style of discussion. I marvelled at those who practiced 'Ordinary Language philosophy', intellectually elegant and academically arrived. Paul despised them. He considered them totally decadent, and completely ignorant.
> [...] Feyerabend – as in the case of logic – took language as not being that important. He thought that by concentrating on language, philosophers would forget about nature, about the world of which we yet know so very little.[38]

These were prejudices shared by a tradition stretching from Mach through Popper, and Feyerabend did not overcome them, in spite of his numerous and interesting debts to Wittgenstein. He could never accept Wittgenstein's suggestions that philosophical problems are sometimes solved, not by bringing forward new experiences or materials or truths, but by rearranging that which has long been known and assumed, so as to display its grammar and symbolic force. Philosophy's battles are waged by means of the very language that may also bewitch our understanding (cf. §109 of the *Investigations*).

That philosophy concerns itself with myth and superstition, and that science in our time has produced new myths with and about symbols, without ever having solved the old ones, is an idea common to Wittgenstein's early and later philosophies, and one that, excepting its concern with investigating language, did have an impact on Feyerabend. It was, after all, from the Austrian poet and literary critic Paul Ernst's discussion of Grimm's *Fairy Tales*, and not from Frege or Russell, that Wittgenstein drew the *Tractatus'* famous remarks about philosophical problems arising from misunderstandings of the logic of our language (*Sprachlogik*).[39]

Yet Wittgenstein's and Ernst's idea seems to have been that myths we have created about and with language are themselves part of our present philosophical reality, part of our very natural history, as distinctive a stylization of nature and ourselves as any arrangement of shells and skeletons, stones and skins, paintings and sculptures and jewelry, to be seen in the halls of a Viennese museum. We must then read the *Investigations* as offering deep criticisms of Feyerabend's (and Mach's and Popper's) dispersion of philosophers' preoccupations with language as hopelessly reductive and uninteresting. See, for example, in Part II, xii (p. 195):

> If the formation of concepts (*Begriffsbildung*) can be explained by facts of nature, should we not be interested, not in grammar, but

rather in that in nature which is the basis of grammar? – Our interest certainly includes the correspondence between concepts and very general facts of nature. (Such facts as mostly do not strike us because of their generality.) But our interest does not fall back upon these possible causes of the formation of concepts; we are not doing natural science; nor yet natural history – since we can also invent fictitious natural history for our purposes.

I am not saying: if such-and-such facts of nature were different people would have different concepts (in the sense of a hypothesis). But: if anyone believes that certain concepts are absolutely the correct ones, and that having different ones would mean not realizing something that we realize – then let him imagine certain very general facts of nature to be different from what we are used to, and the formation of concepts different from the usual ones will become intelligible to him.

Compare a concept with a style of painting. For is even our style of painting arbitrary? Can we choose one at pleasure? (The Egyptian, for instance.) Is it a mere question of pleasing and ugly?

2. Feyerabend on Wittgenstein's Philosophical Investigations: Some Aesthetic and Philosophical Limitations of Feyerabend's discussions

Feyerabend, unlike Wittgenstein, was driven by a preference for dramatic historical narrative. One might even say that his overarching philosophical idea was to replace the idea of scientific method with historical drama. The latter form of writing he did not view as part of the "ordinary", or part of "common sense", but as critical of these. Indeed Feyerabend conceived ideas and concepts to be evolving characters or personalities, best scrutinized by being brought to life on a narrative stage. "Science," as he wrote in his autobiography, "is a story, not a logic problem", a story that is to be performed and animated through narration, not merely described, much less left alone.[40] (We shall return to consider Feyerabend's take on Wittgenstein's remark that "philosophy leaves everything as it is" (*Investigations* §124) in section 4 below.)

Philosophy's aims could be for Feyerabend epistemological or political, or both at once. Dramatic performances of works like Brecht's *Galileo* were exemplars of philosophy of science, containing a reality within them that can shock and change us – not merely through sympathy or identification or catharsis, but through observation and experience, a lived immersion in the enactment of an event. One of the falsities of both traditional and positivistic empiricism, for Feyerabend, lay in their restricted conception, both of experience and of reality: the character of Mother Courage, for Feyerabend, is a part of reality we can learn from, directly observe, and learn to know better. He was and remained a student of the Renaissance: "Giotto, with his determined stylization of events, is one of my favorite artists", he wrote.[41]

By contrast, Wittgenstein spent much of his philosophical life aiming to avoid the directly political and the didactically theatrical. He conceived words and their uses as allegories of our individual lives, and logical features of our language as (shifting) glimpses of particular facial expressions or physiognomies immersed in contexts of use. His recurrent comparisons between logico-grammatical features of our language and physiognomies were intended, not to further a general account of relations between syntax and fact (or facial features and character) so much as to resist such accounts when they overlook the importance of context to structure.

Early, middle and late Wittgenstein begins with an understanding that there is both a gap, as well as a series of important connections, between the particular pictures we form of our characters and ideals (the rules and standards we enunciate and profess allegiance to) and what we actually do in projecting them (how we follow and apply the rules).[42] Philosophy was for him a way to expose the places where vanity, received authority, unclarity and lack of resolve blunted his powers of expression, making him dishonest with others and himself. His difficulties were the kind of difficulties we each face, one by one, as we inherit and pass on a language. Indeed, it is one of Wittgenstein's great intellectual contributions to have made this struggle worthy of the name of philosophy.

As I see it, the centrality of this contribution eluded Feyerabend. This is partly indicated by Feyerabend's aesthetic preferences. The point of Wittgenstein's focus on an individual's inheritance of language was not to defend either relativism or anarchism; the point was not any "ism" at all, but instead, something akin to the idea that to imagine and confront a language (or a concept) is to imagine a form of life, either a particular individual life form or a way of living we imagine for ourselves. Wittgenstein and Feyerabend agree that dogmatisms finding a home in the everyday can be expressed and addressed philosophically through the refashioning of belief, either in a drama or in an imagined language-game: here we may see jejune, platitudinous phrases and ideas thrown into new, unfamiliar contexts and allow ourselves to be shocked and provoked, until at last the ideas, and our uses of them, become objects of criticism. But Wittgenstein's "album" of glimpses and pictures and competing models does not fit very well with Feyerabend's Giottoesque preference for didactic dramatic narrative.

The socio-economic, personal and cultural backgrounds of these two Viennese philosophers were quite different, creating very different spaces within which they philosophized. This is illustrated by Feyerabend's account, in his autobiography, of an early visit he paid to the Wittgenstein family residence in Vienna in 1950. Elizabeth Anscombe, who had come to Vienna to work on her German, had tried to explain Wittgenstein to members of the Kraft Circle, "without much success", as Feyerabend records.

> Hearing of our reaction, Elizabeth suggested that I approach Ludwig Wittgenstein himself, who happened to be in Vienna. I went to the Wittgenstein family residence (not the house in the Kundmanngasse). The entrance hall was large and dark, with black statues in niches all over the place. "What do you want?" a disembodied voice asked. I explained that I had come to see Herr Wittgenstein and to invite him to our circle. There was a long silence. Then the voice – the housekeeper, who spoke from a small and almost invisible window high up in the lobby – returned: "Herr Wittgenstein has heard of you, but he cannot help you."[43]

This is pure theatre – by which I do not mean to question its veracity, but to describe its literary, nearly cinematographic, ambition and power. It illustrates, minimally, one reason Wittgenstein could not have resided in Vienna to practice philosophy. The prospect of finding himself constantly subjected to the humiliating image of being nothing more than a wealthy amateur, cosseted by a prominent, never fully assimilated family of parents, sisters and brothers living in a palace (never mind the way the family history had unfolded after the *Anschluss*[44]) clearly made the prospect of a life away from the limelight of social scrutiny impossible for him to imagine in Vienna where it was possible to imagine in Cambridge. Ludwig could barely make a move without his family being an issue.

Feyerabend, by contrast, did not suffer in that particular way. His family was, socially speaking, relatively on the margin and private. He was an only child. There were consequences of this. Though each philosopher faced suicide in his immediate family, it was Feyerabend, not Wittgenstein, who was able to write about this darkness publicly, in his autobiography. The opening chapter of *Killing Time*, where he treats of his mother's taking of her life, is as moving a meditation on such a thing as any philosopher I know has written. It is dramatic narrative of a high order.[45]

Feyerabend did meet Wittgenstein, exactly once, on what was to be Wittgenstein's final visit to Vienna. After Feyerabend's aborted visit to the Alleegasse, Anscombe suggested that Feyerabend try writing a letter to Wittgenstein, but without making it "too subservient".

> I wrote roughly as follows: "We are a group of students, we are discussing basic statements, and we are stuck; we hear you are in town – perhaps you can help us." Wittgenstein seemed to like what I had written. "I've received a rather nice letter," he said, again according to Elizabeth, emphasizing "rather," and was thinking of coming. Now the science students balked. "Who is this guy?" they asked, "and why should we listen to him? Anscombe was bad enough!" I calmed them down and reserved a room. On the day of the meeting I had a cold. Being rather ignorant in medical matters I swallowed tons of sulfonamides and sat down in eager anticipation. The hour arrived. Kraft was there, the philosophers were there – but no Wittgenstein. Afterward, Elizabeth told me how difficult it had been for Wittgenstein to negotiate this particular event. Should he come at the correct time, sit down, and just listen? Should he come a little late and enter with a flourish? Should he come very late, simply walk in, and sit down as if nothing had happened? Should he come very late and make a joke? At any rate, I started summarizing what we had been doing and made some suggestions of my own. Wittgenstein was over an hour late. "His face looks like a dried apple," I thought, and continued talking. Wittgenstein sat down, listened for a few minutes, and then interrupted: "Halt, so geht das nicht!" ("Stop, that's not the way it is!"). He discussed in detail what one sees when looking through a microscope – these are the matters that count, he seemed to say, not abstract considerations about the relation of "basic statements" to "theories." I remember the precise way in which he pronounced the word *Mikroskopp*. There were interruptions, impudent questions. Wittgenstein was not disturbed. He obviously preferred our disrespectful attitude to the fawning admiration he encountered elsewhere. Next day I was in bed with jaundice – the sulfonamides had done me in. But Wittgenstein, I heard, had enjoyed himself.[46]

Here, once again, is autobiography as character theatre. Above and beyond his report of Wittgenstein's revulsion at being treated with the wrong sort of deference, there is the interesting fact that Feyerabend recalls Wittgenstein having addressed what was to become the central philosophical theme of Feyerabend's own writing, viz., the lack of a sharp distinction between observation and theory. The limitations that plagued Feyerabend's appropriation of this theme, some of which I have already enumerated, are already revealed in this quote. For the contrast Feyerabend puts into Wittgenstein's mouth is between concrete perceptual activity and abstract theory of language (or of theory), and that contrast sounds a false note about Wittgenstein's philosophy.

Feyerabend's involvement with Wittgenstein began in Vienna, but intensified with his move to England in 1952. He arrived first at the London School of Economics: having just missed the chance to study with Wittgenstein, who died in 1951, he went to Popper instead. As soon as he settled in London, he set to work reading Wittgenstein's later writings. He found them difficult:

> Wittgenstein resisted in a different way [from Von Neumann]. His writings sounded like fragments of a novel, but it was not clear who the actors were and what their actions meant.[47]

The unclarity of actors, identities, and meaning confronting the reader of *Philosophical Investigations*' opening – the literary forms that lead readers to ask questions such as, "Where are we going here?", "Are these arguments?" "Who is speaking where?" "What does this author think?", "Where, if ever, do explanations end?" – are more or less familiar to its readers. On the opening stage of the book we are confronted with a multilogue of voices, intertwined yet distinguishable, one and many, with no clear place of habitation, no dramatic *personae*, no specific history, no clear motivation, and no clear future, common or otherwise.

And yet for this very reason it is difficult to imagine trying to experience the *Investigations*, as Feyerabend did, as novelistic. To have a narrative is to have a story with characters and a plot, a string of themes, contexts, and voices that cohere. Pretty obviously the *Investigations* not only lacks these elements, but deflects and defeats their arrangement and imposition upon its surface. The limitations of imposing on the *Investigations* one overarching narrative structure or philosophical point are, by now, well known among its readers, if only because some of the most widely circulated initial narrative accounts of the work failed so badly to do justice to readers' sense of the text's difficulties, complexities, and philosophical possibilities. (Two narratives less likely to appeal to readers now than in the late 1950s are "the *Investigations* is essentially a correction of views held earlier in the *Tractatus*" and "the point of the *Investigations* is that all language is guided by rules". It is to Feyerabend's credit that he resists these reductive interpretations of the text.) The best I can muster, if I try to imagine a novelistic form for the *Investigations*, is to picture an epistolary form – but that simply makes it confessional, as Wittgenstein evidently suggests that it is in quoting from Augustine. Such writing is precisely not narrative (however dramatic and theatrical it shows itself to be), even in what Feyerabend calls a "fragmentary" sense. It is nothing at all like *Mother Courage*, Feyerabend's ideal of a good play.

Just a few years after Feyerabend's review of the *Investigations* was published, Stanley Cavell began to lecture at Berkeley on the philosophical significance of the *Investigations'* literary form. On Cavell's view, the text is a characteristically modernist work, analogous in its effects and preoccupations to works of Beckett such as *Endgame* and *Waiting for Godot*.[48] As Cavell sees them, works such as these confront us with an immediate present whose conditions, specific locations, assumptions and applications are to be reconstructed by the reader.

The one remark Feyerabend makes about *Waiting for Godot* in his autobiography is that he "thought it was the pits"[49]. Was this a stab at expressing what Cavell called the "hidden literality" of Beckett (exemplified by the "ordinariness" of his characters and the literalness of his truths) and thereby an attempt to undo Beckett's difficult observations (what Cavell calls our "curses") by means of a joke?[50] Or was Feyerabend just too in love with the Giottoesque form of stylized narrative of historical events to grasp Beckett? Was he too in love with this form to grasp the later Wittgenstein?

It seems to me that Feyerabend's lifelong obsession with undercutting, by means of dramatic historical narrative, the notion of an isolated, purely observational experience unadulterated by conceptual or theoretical baggage contrasts with Cavell's remark that Wittgenstein, like Beckett, is driven to expose and enact the attempt to construct such an immediate, purely given, uninterpreted present precisely by depicting its human cost.[51] What Feyerabend seems not to have brought himself to question – unlike Beckett, unlike Wittgenstein, and unlike Cavell – is the idea of a directly verifiable observation of an event in reality given through the lens of an unambiguously structured narrative. This is suggested by Feyerabend's preference for drama in which the conditions informing the means of expression of the play are not put constantly into question by the terms of the play itself. (We find no reference to Shakespeare in Feyerabend's autobiography, despite many anecdotes about his experiences of theatre and opera.[52])

Here are an obvious series of differences between Feyerabend and Wittgenstein. The author of the *Investigations* exploited the broken-off line of thought, the doubled aphorism, the punctuation mark, the contrasting image, and the echo, precisely so as to bring about in his reader reflection upon his or her ways of formulating and interpreting the *Fragestellungen* of philosophical questions.[53] That multiple, possibly competing narratives, pictures, or standards shape our descriptions of our experiences and actions, making them what they are, is an obvious theme of much of Wittgenstein's philosophy. He had already written in the *Tractatus* that logical features of speech are punctuations, contextualized elements more like musical transitions, resolutions, and variations, than the kind of elements expressed by referring structures or individual signs (cf. *Tractatus* 5.4611). All this, I am suggesting, is very unlike Feyerabend's conception of the conceptual and/or grammatical saturability of perception and expression, which was a mixture of deference to empirical psychology and to a rationalistic vision of concepts.

Despite his initial difficulties with Wittgenstein's writing, Feyerabend persisted in reading him:

> Anscombe had earlier given me manuscript materials, including a photocopy of the "Remarks on the Foundations of Mathematics." I found this manuscript extraordinarily exciting but I could not say why. I had also read the *Investigations*. Now that the work was published, I tried to get to the bottom of things. I rewrote the text, turned it into a treatise, and used four different symbols to signal my interventions: normal quotation marks for Wittgenstein's text, crosses for paraphrases, stars for elaborations, and still another type (I have forgotten which) for critical comments. I knew that Wittgenstein did not want to present a theory (of knowledge, or language), and I did not expressly formulate a theory myself. But my arrangement made the text speak like a theory and falsified Wittgenstein's intentions. I gave the essay to Elizabeth [Anscombe] for criticism. She prepared an English version (my text was in German) and sent it to Ryle. He sent it right back: "An efficient condensation, not a review." Malcolm was more yielding. And so the paper monster I had produced for my own enlightenment was published in the *Philosophical Review* of 1955.[54]

What emerged from this initial immersion was his first publication, the only essay Feyerabend ever explicitly devoted to Wittgenstein, and yet at the same time an essay that was to solidify his reputation as a Wittgenstein exegete, for in 1966 it was to be reprinted in the widely-circulated anthology edited by George Pitcher, a collection that set the standard for interpreting Wittgenstein for at least the next twenty-five years.[55]

It is the most deferential essay Feyerabend ever wrote. In it he tries to stick to Wittgenstein *qua* Wittgenstein, to quote copiously and simply "rearrange" the text, to stay as close as he can to description and characterization, rather than interpretation, defense, or explanation. What is fashioned is, however, a specific interpretation of Wittgenstein's thought. And what Feyerabend leaves out of account in the review is as important as what he brings into it.

Feyerabend's treatment of the *Investigations* is not without merit and insight. He spends the bulk of his review exploring Wittgenstein's remarks on rule-following, zeroing in on the kinds of considerations Wittgenstein brings forward to dislodge the philosophical idea that understanding is a mental state. He appreciates that, as he puts it, "Wittgenstein does not want his reader to discover that reading is *not* a mental process. For if 'mental process' is used in a metaphysical way in 'reading is a mental process,' it is used just as metaphysically in 'reading is not a mental process' (cf. [§]116), and his aim is 'to teach you to pass from a piece of disguised nonsense to something that is patent nonsense' ([§]464)".[56] In substance and preoccupation Feyerabend's review reflects the happy influence of its translator into English, G.E.M. Anscombe. Like Anscombe, Feyerabend emphasizes aspects of Wittgenstein's thought that emerged from his reactions to Frege. (Since in the mid-1950s Frege was only beginning to be appreciated properly, both as a philosopher in his own right and as a central influence on Wittgenstein of at least comparable importance to Russell, this is noteworthy.[57]) In addition, Feyerabend nicely emphasizes, as did Anscombe, continuities in the evolution of Wittgenstein's thought (these, as scholars have since come to emphasize, are in certain respects rather deeper and more wide-ranging than has often been supposed). He singles out the notion of *elucidation*, of philosophy as an activity and not a theory, as central, both to the *Tractatus* and to the *Investigations*, and holds that there is one Wittgenstein, for the *Investigations* is "an application of the main ideas of the *Tractatus* to several problems".[58] Finally, Feyerabend is concerned to warn against the mistake of aligning Wittgenstein too closely or coarsely with logical empiricism, as the Vienna Circle had done. Here Feyerabend's insistence on the conceptually saturated character of observation plays a useful role.

Feyerabend leaves to the side, however, what seems equally evident in drawing a contrast between Wittgenstein's later philosophy and any recognizeable form of empiricism. Wittgenstein must be read, not only as a critic of the myth of the uninterpreted sensory given, but equally as a critic of the more or less uncritically rationalistic view of concepts that has been viewed, at various points in the history of philosophy, to be able to refute it. Here Cavell's slightly later anti-empiricist reading of Wittgenstein, which surrendered this rationalism by raising important analogies between Kant's third *Critique* and Wittgenstein, would constitute a decisive advance over both Anscombe and Feyerabend.[59] Though Feyerabend appeared to appreciate the centrality to the *Investigations* of Wittgenstein's discussions of rules and rule-following, he remained unable to entertain the idea that Wittgenstein might best be read as mounting a wide-ranging attack on how myths about concept-possession arise, particularly by exploring how tendentious presuppositions are already embedded in the description of language as a fixed calculus of rules and/or beliefs.

Thus, on the one hand, Feyerabend emphasizes the importance to Wittgenstein of the idea that there is no coherence to the notion of an "ideal" language, that upon scrutiny the idea falls apart (or, as Feyerabend nicely puts it, "becomes obsolete").[60] But on the other hand he holds on to the Popperian conviction that the distinction between so-called "ideal" language philosophy and so-called "ordinary language" philosophy is a ground level, obvious distinction of method based on subject matter. (It is not.) He sees that Wittgenstein's attack on mentalism (on the idea that reading and understanding are mental processes of some unspecified kind) is not, or not precisely, to be understood as a discovery or theoretical result – thereby transcending the limitations of intellectual fashion of the time, which held that Wittgenstein must be unproblematically read as a logical behaviorist. At the same time, he fails to see that this insight can only be internalized properly if it forms part of a careful investigation of conceptions of meaning, rather than an insistence on surrendering the investigation of language.

Feyerabend was always torn between feeling that scrutiny of language must trivialize philosophy, and so should be irrelevant to its aims, and having strong opinions about how philosophers should and should not investigate language. While he expressed in his later life an admirable unwillingness to define his own position in terms of an "ism", in this early work we see him mired (as he remained mired through at least the late 1970s) in this tendency. His insights into the *Philosophical Investigations* were then only partially grasped, skewed by a failure to reflect on the difficulties and complexities facing any philosopher who would try to convey them without reflecting on how various pictures of meaning and concept-possession might block their philosophical availability. This goes to show that the position that one has no position is not always to be trusted; or, more generally, that what a philosopher says about what, in the main, he or she is doing, is not always the same as what he or she actually does.

The limitations displayed in Feyerabend's early presentation of Wittgenstein are, in these respects, glaring. He does not emphasize the thoroughly self-reflexive, dialectical character of Wittgenstein's writing, the way it is always rounding on its own interlocutory voices and attempting to depict philosophical exchange rather than further it. This leads him to downplay Wittgenstein's determination to scrutinize the philosophical implications of the idea that there is in the end nothing general to be said in philosophy about anything in advance of what Wittgenstein calls an "investigation" of our concepts. Feyerabend thereby gives insufficient attention to the idea that in the very effort to extricate oneself from a particular conceptual problem or error one may produce new, unfamiliar, and possibly more obscure forms of confusion – which implies that we do well to investigate the sources of philosophical illusion. Instead he chides Wittgenstein for treating this as a "simple" matter, objecting that "there are philosophical systems, philosophical theories; and it needs to be explained how it is that they come into existence if 'nothing is concealed'"[61]. This leaves out of account one of Wittgenstein's most important starting and ending points, namely, that what is least obvious, and sometimes most philosophically important, is that which we take to be obvious or trivial or *a priori*, the assumptions we take for granted at the outset, before a position is articulated. The distinctions between the philosophical and the everyday, between ideal and ordinary language, are to be investigated, not invoked as starting points. They do not define any kind of method, or differing subject matters for the philosopher.

I find it then a pity that Feyerabend accedes in the mistaken idea that Wittgenstein's remarks evince little beyond constructivism and/or verificationism, that the *Tractatus* and the *Investigations* defend the view that "seeing the world rightly means playing the language-games without being troubled by philosophical *questions* or by philosophical *problems*", and that there is for the later Wittgenstein, as Feyerabend puts it, "a definite boundary between the card-houses of philosophy and the solid ground of everyday language".[62] These misreadings are, though familiar, much too simplistic. From Feyerabend's point of view, of course, they were ways of protecting Wittgenstein from what Feyerabend himself regarded as Wittgenstein's darker side, the conservative defender of "ordinary language" and "common sense", with all their incipient bestiality and criminality.

3. Feyerabend's Uncritical Rationalism

I have said that Feyerabend missed Wittgenstein's depth partly because he inherited from early analytic philosophy a highly rationalistic picture of language-use and concept-possession, the idea that logical analysis had to be in the business of offering necessary and sufficient conditions in the traditional sense. Though one might trace this philosophy of concepts back to Leibniz, in the twentieth century it often went hand in hand with the (unLeibnizean) idea that logic is something trivial and empty, an idea that Feyerabend, as we have seen, held.[63] On this kind of view, if you find out that something is a mark of a concept that wasn't conceived as a mark of it before, then the meaning of the original concept goes to pieces, and you are stuck with a change to a completely new concept, with no method or rule to get you from one to the next. This rationalistic, overly wooden picture of the conceptual saturation of experience is precisely what led Feyerabend to embrace and emphasize a picture of scientific progress as arational if not irreducibly irrational, especially in revolutionary contexts.

In his youth Feyerabend was attracted to Carnap's idea of a purification of scientific language, but not very fervently or thoughtfully. In the one essay he ever wrote on the the nature of analysis, Feyerabend is surprisingly simplistic, unfazed by evident difficulties with the whole idea that were being raised by many philosophers in the 1950s – most famously by Quine, but also, evidently, by Wittgenstein as well.[64] As we have seen, Feyerabend always considered logical analysis of any kind an empty, purely formal pursuit, and certainly a waste of time for philosophers.

By the late 1950s Feyerabend had moved beyond exposition of Wittgenstein. In this phase, he came to defend what he called the "contextual theory of meaning", a view he explicitly claimed to derive from Wittgenstein.[65] On this kind of view, a sharp dichotomy between theoretical and observational terms is undercut by insisting that the meaning of *all* terms in the language is gained through their logical place within the theory. This view of meaning vastly oversimplifies and vastly underestimates the force of two recognizeably Wittgensteinian ideas. First, there is the relativity and variety of kinds of words and of categorical distinctions: an account of meaning ought to be contextually sensitive and ought not to be monolithic across all words of the language. Second, it is a mistake to assume that perceptual reports must always be inferentially located in a single, closed system or language in order to establish the relevance of conceptualization to their effects.

Of course, Feyerabend was not alone in failing to grasp the significance of these two points. One excuse for his naiveté might be that it seems never to have been very easy for *any* analytic philosopher to work out a stable combination of these two Wittgensteinian ideas. Compare, for example, Quine. It took Quine, that latter-day empiricist, a surprisingly long time to assimilate these Wittgensteinian insights into his system in such a way as to address worries one might have about the Kuhnian (or Feyerabendian) "theory-ladenness" of observational terms and/or sentences. He did so only in the 1980s, by introducing a distinction between treating a sentence "holophrastically" and treating it "analytically".[66] (Quine was explicit that this was his way of accommodating Kuhn's and Feyerabend's talk of the theory-ladenness of observation terms.[67]) One reason it took Quine so long to settle on a distinction that would respond to Kuhn and Feyerabend is that he had always been – ironically – an empiricist who worked with no general notion of *experience*. If one is an empiricist like this, one need not ponder the question whether all experience is or is not conceptually saturated. Another reason was Quine's unclarity about how best to view our language, a conceptual scheme in which we begin, end, and naturally acquiesce, as a systematic "theory of the world".

Now it has never been easy for Quine's readers to see how Quine can avoid relativism and/or instrumentalism, given his holism, so Feyerabend is not alone in having fallen into this quandary.[68] But the way out of the quandary lies in Quine's refusal to employ a notion of meaning across the board, for all terms of the language at once (hence, within science as he conceives of it). Quine was instinctively far more thoroughly anti-rationalist than was Feyerabend. He appreciated how deeply embedded in rationalism were the dogmas that informed traditional empiricism. In Quine's story of the progress of empiricism, the "shift of attention from ideas to words" (a shift Quine credited to John Horne Tooke, a follower of Locke) still retained within it the empiricist tendency to suppose that to the impression or observation or element of experience corresponded (directly, immediately, and in some kind of isomorphic way) the word, the meaning, a unit of theoretical significance.[69] Like Austin and Wittgenstein, Quine respected the difficulties of surmounting this tendency. *Sense and Sensibilia* and "Other Minds" are, for their part, as much designed to expose the pseudo-rationalism of so-called empiricists as they are to attack the concept of an unconceptualized sense datum.

Obviously Quine and Austin lacked immersion in the Nietzschean, Splenglerian view of culture I have ascribed to Wittgenstein and Feyerabend. Their respective styles and manners of writing, though each is unquestionably a master of the craft in English, are hardly Austrian. It would be better to say, however, that neither Quine nor Austin had a religious bone in his body than to say that either one converted science into a religion. Neither felt a need even to bother with atheism and secularism, as had Voltaire. Quine's specific contribution here is comparable to Hume's in an earlier century: wittingly or unwittingly, Quine purged empiricism and pragmatism of their false scientism precisely by exploring their scientific limitations. A broadly Darwinean starting point in thinking about the progress of human culture (and language as part and parcel of it) was for Quine not even something to be argued for, it was simply what every minimally educated person in our time may be assumed to have learned at a mother's or a father's knee. For this reason – a theme I shall return to toward the end of this essay – Quine and Wittgenstein and Austin rested no arguments on any general concepts of "science" or "knowledge" or "certainty".[70]

In introducing a notion of "incommensurability" in his 1962 essay "Explanation, Reduction, and Empiricism", Feyerabend focussed on the limitations of Hempel's hypothetico-deductive model applied to "universal theories", large-scale theoretical structures such as Aristotelian dynamics or Newtonian mechanics. But he slid from there to treating all observation terms as theoretical terms in the same monolithic way. As the later Wittgenstein would have emphasized, Feyerabend's slide distorts our talk about perception. For one thing, language as such is not – or anyway, not obviously – a theory. For another, to insist on a picture of the conceptually-saturated character of all experience presupposes that we are clear on what we understand, both by "experience" and by "conceptually saturated" or "theory laden". Here Feyerabend was just not clear.

Indeed, it was a main thrust of Putnam's semantical externalism, enunciated in the mid-1960s, to question, both this oversimplified picture of concept possession and the (connected) verificationist tendency to rely uncritically upon the notion of meaning in philosophical debate.

The central thrust of Putnam's semantical externalism is negative, and it is directed, not accidentally, at Feyerabend himself. For it was coming to ponder his own remarks about the notion of a "semantical rule" in his debate with Feyerabend that set Putnam on the path toward articulating his externalism.[71] Part of the reason for the *succès fou* of Putnam's 1975 externalist paper "The Meaning of 'Meaning'" is that it seemed to provide a way out of the incommensurability (and the apparent relativism) that Feyerabend and Kuhn appeared to have been defending.[72] Thus it is that Feyerabend made an indirect contribution to so-called philosophy of language, all the while denying that the notion of meaning or the examination of language was important for philosophy.

On the one hand, Feyerabend protested vigorously against the theory of meaning and all forms of linguistic analysis as hopelessly irrelevant to philosophy. On the other hand, he was mired in preconceptions about meaning that philosophers like Putnam were attempting to work beyond. Putnam makes the series of (related) points in this way:

> One might, of course, take the radical line that *any* change in theory is a change in the meaning of terms. (Which term? All of them? Even the logical connectives? Why not?) But I expect Feyerabend would not want to take this line. For to say that *any* change in our empirical beliefs about Xs is a change in the meaning of the term X would be to *abandon* the distinction between questions of meaning and questions of fact. To say that the semantical rules of English cannot at all be distinguished from the empirical beliefs of English speakers would just be to throw the notion of a semantical rule of English overboard.
>
> What is curious is that Feyerabend does *not* follow this course. Indeed, many of his purposes would have been better served had he chosen to follow Quine in repudiating the theory of meaning altogether. He wishes to show that linguistic philosophy is irrelevant and misguided; clearly, if it all rests on a mistaken notion, that there are such things as *rules of language*, then it is badly misguided [...]
>
> ... [Feyerabend also] wishes to show that false theories are presupposed by ordinary language; [but] if no distinction is to be drawn between 'ordinary language' and 'common sense' (i.e. the everyday beliefs of most speakers), then this is just to say that most people believe many false things, and who has ever doubted *this*?
>
> ... [But] one of Feyerabend's main conclusions ... is that, since ordinary language presupposes false theories, we may have to *discard ordinary language*. It is no surprise that we may have to

change many of our common sense beliefs; we may have to say different things *in* ordinary English, French, German, etc.; these beliefs are the 'meaning'; only by supposing that they are also the 'meaning' in the customary sense can one obtain the conclusion that we may have to discard everyday English, French, German, etc., altogether. Here Quine and Feyerabend would, of course, part company. If the 'ordinary usage' of a term X depends on the false proposition that all X are P, then Quine would recommend that we start saying instead what is true, namely 'some X are not P', and not that we necessarily drop the term X. If 'common sense' is false, let us find out what is true. Many technical terms may have to be introduced; but that ordinary non-technical language cannot be used to say what is true, that it is somehow *essentially infected with falsity* is a conclusion that Quine would reject as he rejects 'essentialism' in general. Indeed, Feyerabend's views are an unholy mixture of Quine's refusal to separate meaning from empirical theory and just the essentialism that Quine attacks.[73]

Putnam was zeroing in on two main points. First, holism about the theory-ladenness of observations without critical reflection on the notions of meaning, belief and language cannot work. Differently put, an abandonment of the analytic/synthetic distinction as a dogma is laudable, but it does not free philosophers from the burden of investigating the notions of meaning and empirical content along far more complicated lines than Frege, Russell, Feyerabend, and others had envisioned. Second (an application of the first point), a global distinction between so-called "ordinary" language and so-called "scientific" language cannot be uncritically relied upon. In particular the distinction, much touted by Popperians and taken over by Feyerabend, between so-called "ordinary" and so-called "ideal" language philosophy is miscast. First, it is obvious that mathematics, including formal logic as an emerging branch of it, is part of the ordinary insofar as it grows (as it does grow) from our working uses of language. (Science, as Quine laudably put it in the first chapter of *Word and Object*, is just a self-conscious form of common sense.) Second, since our descriptions of ordinary speech acts and everyday, recognized procedures with certain of our verbs have often, in variously documentable ways, been skewed by the philosophical effects of formal logic as a guiding philosophical touchstone or organon, there is no impermeable barrier we may rely on to inoculate one form or area of language from another (this is a lesson to be learned from Austin and from Wittgenstein). Given the philosophical history of the last

150 years, we ought to see that this is partly accounted for by the fact that at least some of the diseases a formalism enshrines it has inherited from preconceptions about the uses of ordinary language which it is being taken to model.[74]

However we are to understand them, distinctions between the ordinary and the philosophical are not to be equated with a distinction between ordinary and formalized languages, nor do "ordinary" and "ideal" languages form different subject matters (or methods) for philosophers that will necessarily yield right or wrong answers. It does not make clear sense, given where we stand at this point in the evolution of human culture, to take scientific belief to be something over and apart from ordinary belief, or something called "common sense". The distinction between "theory" and "practice" will not, in particular, support these distinctions.

It follows that philosophers who illuminate for us the complexities of such distinctions (like Wittgenstein, Austin, and Quine) should not be read as exclusively tracking one sort of distinction, method, or subject matter. We do not have to do here with rival methodologies or subject matters for philosophy at all, much less with a global contrast between "science" and "common sense". And if this is right, then it is surely a significant underdescription of the philosophical terrain in which we find ourselves to label Wittgenstein and Austin "ordinary language philosophers" as opposed to "scientific" philosophers like Quine and Feyerabend. This is an error into which Feyerabend tended, at least on occasion, to fall.

4. Feyerabend on Quine and Austin

Feyerabend misunderstood, or only partially understood, both Wittgenstein's motto to the *Investigations* from Nestroy ("progress has this in it, that it always appears greater than it is") and his remark that "philosophy leaves everything as it is" (*Investigations* §124).[75] The former does not deny the concept of progress to philosophy, but warns only against certain misconceptions about what it must look like.[76] The latter warns against taking what Wittgenstein writes either as a direct contribution to the foundations of science or as an anti-scientific critique. It also raises an explicit question about the traditional philosopher's wish to talk about "everything", a form of talk connected in the surrounding passages with false foundational goals.[77] That it might be difficult to leave talk about everything in the world, in one's language, in science, or in mathematics as a whole "alone", that the ability to do so is in and of itself a form of philosophical achievement was an idea, as I have argued, that Feyerabend never fully digested. Neither of Wittgenstein's remarks, in the context of the *Investigations*, function as flat assertions reflecting a bias in favor of something called "ordinary language". Instead, they are clues, puzzles to be worked out in the course of the text. They point us toward problems about the grammar of our talk, not solutions. But for Feyerabend, as for many readers of the *Investigations*, they professed a kind of flat conservatism.[78] Feyerabend wanted to attack "common sense" and "science" as categories, not investigate what talk about everything might or might not amount to.

This may be seen by considering a passage in Feyerabend's autobiography where he appeals to Nestroy to attack Quine and Austin. Indirectly, the passage as a whole is an allusion to Wittgenstein's motto to the *Investigations*. Since many of the themes on which I have already touched are in evidence in this passage, I want to consider it at some length.

> Even as a student I had mocked the intellectual tumors grown by philosophers. I had lost patience when a debate about scientific achievements was interrupted by an attempt to "clarify," where clarification meant translation into some form of pidgin logic. "You are like medieval scholars," I had objected; "they didn't understand anything unless it was translated into Latin." My doubts increased when a reference to logic was used not just to clarify but to evade scientific problems. "We are making a logical point," the philosophers would say when the distance between their principles and the real world became rather obvious. Compared with such doubletalk, Quine's "Two Dogmas of Empiricism" was like a breath of fresh air. J.L. Austin, whom I heard in Berkeley, dissolved "philosophy" in a different way. His lectures (later published as *Sense and Sensibilia*) were simple, but quite effective. Using Ayer's *Foundations of Empirical Knowledge*, Austin invited us to read the text literally, to really pay attention to the printed words. This we did. And statements that had seemed obvious and even profound suddenly ceased to make sense. We also realized that ordinary ways of talking were more flexible and more subtle than their philosophical replacements. So there were now two types of tumors to be removed – philosophy of science and general philosophy (ethics, epistemology, etc.) – and two areas of human activity that could survive without them – science and common sense.

But that was not the end of the story. Science and common sense are not as simple, self-contained, and faultless as the critics of their philosophical superstructures, myself included, were assuming. There is not one common sense, there are many (I argued this point with Austin but could not convince him). Nor is there one way of knowing, science; there are many such ways ...

I greatly admire Johann Nestroy, the nineteenth-century Austrian writer of dialect comedies. His couplets, dialogues, monologues, and plays are about perfectly ordinary situations, but he presents them slightly off center. That causes laughter – another rather ordinary, "normal" occurrence. What I have found interesting in Nestroy is that the combination of normalities gradually assumes a sinister complexion. Nestroy uses extremely simple means (such as changes from dialect to high German and back to dialect again) to reveal pretense, deception, and, maybe, a basic crookedness of the entire world. I have read almost all of Nestroy's plays, not once but often, and I have seen many performances. I always felt that I was witnessing a very special phenomenon that might also be applied to scientific jargon. Karl Kraus used the phenomenon to show the incipient bestiality behind an advertisement, a newspaper article, a piece of profound reasoning. Like Austin he invited people to read literally, word for word, what was before them. Unlike Austin he found inhumanity, not just nonsense.[79]

Here Feyerabend fails to appreciate how true anti-method works. Let us reconsider the supposed distinction between "science" and "common sense", using Quine and Austin as our stalking horses. For Feyerabend's misreading of Quine and Austin here is interesting.

Quine practiced anti-method in a form that I admire. One reason this is easily missed is that Quine (something like Wittgenstein, and wholly unlike Austin) periodically offers what appear to be general characterizations of the nature and subject matter of philosophy. However, it is important to understand (just as it is with Wittgenstein) that these characterizations are not what they can appear to be. They are not definitions, restrictions, or *a priori* specifications representing a philosophical opinion. Instead, they present open-ended, flexible metaphors, i.e., opportunities for reflection. They do not represent an effort to pin down a proper method or subject matter for philosophy, but reject the idea of such a project as wrongheaded. Quine did remark that "philosophy of science is philosophy enough", but the context of this remark is often missed. Missed, among many other things, is Quine's idea that interest in the kinds of explications formal logic has on offer is a matter of degree, extent of individual interest, and intellectual focus, not of principle. Missed also is Quine's pluralism: he insists that because there is no severing the relation between so-called ordinary language and the formalized structures of logic, there is room for a plurality of ways of investigating language.[80] Quine, every bit as much as Austin and Wittgenstein, conceived the notion of an "ideal language" to be incoherent.

More deeply, I would emphasize that one cannot begin to understand the contributions of Quine if one thinks that his claim that philosophy and science are continuous with one another is easy to understand. One must ask oneself, and Quine, What does one mean by "science"? Quine makes various remarks in various places about this; it is clear that his conception of science is extremely broad – so broad, in fact, that in the end it becomes difficult to know what kind of content such a general claim really has, even for him (he includes history, for example, in science). He is quite explicit that there is no such thing as a "scientific method", and that the goal of science, like the goal of philosophy, is understanding (not evolutionary survival, truth in the long run, or pragmatic usefulness).[81] Moreover, it is part of his point, in an age of increasing specialization, both to plead for philosophy's role in shaping a loose-knit, wide-ranging, reflective vision of large portions of our culture and to deny that philosophy can achieve a standpoint or method (transcendental or otherwise) that will yield special insights into these.[82] The wile and wit of Quine is to come up with pithy and funny remarks: vivid, memorable, and never carrying the kind of weight one may think, on first or second or third reading, they can carry. As a result, he is constantly misread and misinterpreted.

At his best, Feyerabend too had, as I read him, very little interest in attempting to *define* what philosophy is: in his final interview he reiterated that he "told stories" rather than creating a systematic account.[83] In general there are relatively few positive remarks in his writings about what philosophy is and ought to be.[84] That, I think, might be viewed as a great strength of his writing, a strength that it shares with Austin's. Unlike Quine or Wittgenstein, he never offers us beautiful metaphors or images or allegories of the nature of philosophy as such. Unlike them, he does not offer up quips that are easily misread as restricting or paring down the subject matter of philosophy. As a result, he is much easier to read, and much less misleading about himself, than are Quine and Wittgenstein. But he did less for philosophy. Feyerabend focussed on the observation/theory distinction and on pluralism of approach in the sciences, whereas Austin, Quine and Wittgenstein focussed on pluralism of a different kind, precisely by turning their readers back upon the terms they use to formulate philosophical questions, and confronting them with the complexity, difficulty and breadth of the field of discussion that would ensue if they did so. The point for these philosophers was not to define philosophy, but to look at how the philosophical *Fragestellungen* we inherit define *us*. Ultimately this is a much more complicated, fruitful, and difficult thing to do than Feyerabend seems to have appreciated.

To be truly anti-method is to give up on philosophy as a purveyor of special truths or ideologies, to give up on characterizing its specialness by attempting to gain a point of view that is *sui generis* (whether it is labeled "transcendental", "phenomenological", "common sensical", "non-scientific" or otherwise). It is not simply to reject a particular model of science or even, more generally, the whole idea of a scientific method or science as given by a set of standardized rules and methods of analysis. It is not a free-standing claim about the disunity of science either. More deeply, anti-method in philosophy is about refusing to begin by trying to construct a proper method for, or characterization of, philosophy, because we cannot ensure philosophy's progress by attempting to get either our method and/or our subject matter right.

Now it might be objected that *au pied de la lettre* – and unlike Austin and Wittgenstein – Quine spends a lot of time defending "isms" and philosophical positions, positions such as "naturalism", "empiricism" and "pragmatism". But this, I believe, is misleading. It is not quite what it looks like at first or second glance, it is not a matter of Quine discussing right and wrong, it is instead a matter of his trying to inculcate in his readers judgment, understanding. Quine illustrates beautifully that what a philosopher says about what he is doing is not always quite the same as what he is actually doing.

The notion that a philosopher's job is to defend one or another received view (realism or idealism or pragmatism), the idea that in philosophy the "isms" are what primarily matter, was anathema to Quine. Careful reading of his works reveals that he rests very little on these labels; they are shorthand for starting points he does not argue for, or helpful pointers, reader's orientation for setting up an attack he plans to mount on an oversimplified distinction. The "isms" are never the be-all and end-all of his work, in fact they are most often the object of his recastings and reinterpretations.[85] The art of philosophy, as Quine, Austin and Wittgenstein practiced it, requires judgment in how to draw distinctions that matter, just sharply enough to illuminate a problem, and never more sharply than that. Never so sharply, in particular, that the distinctions or "isms" themselves become needless objects of philosophical obsession.[86] Even when Quine invokes naturalism or pragmatism, these notions carry no more weight than his surrounding philosophy gives them. They are not part of any procrustean ideological bed; he simply did not believe in philosophical movements.

I take these philosophers' love of the vernacular, of everyday language, to have been part and parcel of this attitude.[87] It is intrinsic to their conception of free intellectual inquiry, free expression, and individualism. It is also an index of their unwillingness to take a stand, as did Feyerabend, on the ultimate bestiality and criminality of the everyday world. They had no need for a stance from which to criticize common sense, for they did not take such a category for granted in the first place.

This is illustrated by Quine's brief memorial essay "On Austin's Method", where Quine describes Austin's procedures – not without irony, but with the best form of charity he knew how to muster – as a contribution to science.

> Once there were but a handful of therapeutic positivists and a multitude of chronic metaphysicians. Now there are therapists in every college. The epidemic has been stemmed and the therapy is routine. How are the veteran therapists hereafter to occupy their minds? One way is by directing their efforts against a continuing but less virulent form of the infection, namely, against philosophical perplexity in the lay mind. Ryle in his *Dilemmas* had a successful go at this. Another way is by continuing the kind of language study that went into the therapy, but continuing it now as a line of pure research. Characteristic writings of Austin's seem to fit in here.
>
> Austin's technique, as Urmson has described it in this symposium, is a mode of introspective inquiry into semantics, conducted by native speakers in groups. It is an inquiry that is continuous with portions of linguistics, and probably capable both of benefiting from professional work in that field and of supplementing it. Despite its philosophical antecedents, it is an inquiry whose affinities in linguistics are not in theoretical linguistics; they are in lexicography. It is an inquiry into subtle differences in the semantics, or circumstances of use, of selected English phrases.[88]

Here is a particular vision – one that Feyerabend, as we shall see, appears to have been able to share – of Austin's contributions. I do not say that I agree with it as an account of what makes Austin's work valuable. But I do not suppose that Quine's account of Austin's contributions as part of lexicography would have been wholly foreign to Austin's view of himself either.

Just as he distinguished Wittgenstein from the Wittgensteinians, Feyerabend distinguished Austin from ordinary-language philosophers. Here is Fischer's illuminating report:

> Paul despised them [ordinary language philosophers]. He considered them totally decadent, and completely ignorant. And yet their best and most famous representative hit him – so to speak – intellectually on the nose, and drew blood. John L. Austin, the professor from Oxford who visited Berkeley for one semester, was still more imaginative, and still faster in argumentation than Paul. And he had certainly more practice in academic discussion. Paul changed his mind, and said that Austin was different than the others, the other "Oxford-philosophers", and that he was really a scientist. And with that judgment, I think, Austin would not have disagreed. Austin thought that in some respect what he had been doing was only making a proposal for a new scientific linguistics.[89]

Austin, as Fischer observed Feyerabend having seen, was a master of dialectic. He appreciated the logic of his situation. By refusing to truck in any general notion of meaning, and sticking to example after example of usage, Austin could unmask the errors, superstitions, and fogginess of what he called, on at least one occasion, "the cackle". In the end, Austin was a deep philosopher – and perhaps as deep or deeper than Wittgenstein – if we may allow ourselves to ask, in a wholly un-Austian way, whether the deepest thing of all may be the refusal to work with notions like *deep*. But if he was, it was not because of what he took to be his own overarching aim, not because his contributions are a contribution to philology or lexicography, and not because of a newfangled method or a subject matter to which he pointed our way. Austin's contributions lay in his mastery of the concrete case, in the ways he devised of showing the epistemologist and the ethicist something about his or her own attitude toward various complexities and parallels in our uses of specific parts of our language, for example, in our everyday uses of the verbs "to know" and "to promise" and "to see". Austin's corpus raises the useful question whether, in the end, this is all one *can* do; perhaps one should learn to shift the burden of proof, let one's investigations stand and refuse to try to characterize what, in general, one is doing. Quine used the occasion of his essay on Austin to rail against those who would obsess with the question, What is philosophy?

Let me broach next the utterly boring question, as Urmson called it, of how to classify Austin's introspective semantics. Is it to be called philosophy? To call it that does not, from Austin's point of view as described by Urmson, say much about it; philosophy is "a heterogeneous set of enquiries." I applaud this casual attitude toward the demarcation of disciplines. Names of disciplines should be seen only as technical aids in the organization of curricula and libraries; a scholar is better known by the individuality of his problems than by the name of his discipline. If deans and librarians class some of his problems as philosophical, that is no reason for him to be concerned with other problems that they class as philosophical: his further concerns might just as well be problems that are classed as linguistic or mathematical ...

Does calling Austin's distinctive activity philosophical say *anything* about it? The one salient trait of philosophical inquiries, according to Austin as represented by Urmson, is that for want of standard methods they have not yet hived off under some special name. This criterion is not helpful. The want of standard methods in Austin's work is surely not so dire as to prevent its hiving off under the special name of linguistics.

Actually Austin's work has a genuine tie to philosophy, in a more substantial sense than just what hasn't hived off. It comes in his choice of idioms for analysis. He was no Baconian inductivist, amassing random samples of the world or of the dictionary and scanning them with untendentious eye for unpreconceived uniformities. The *arrière pensée* of *How to Do Things with Words* emerges toward the end of that book: it is "an inclination to play Old Harry with ... (1) the true/false fetish, (2) the value/fact fetish" ...

Historians of science tell us that science forges ahead not by an indiscriminate Baconian inductivism but by pursuing preconceptions, even mistaken ones. I see in Austin's work this kind of progress.[90]

The last paragraph of this quotation is indicative of more than just Quine's complimentary concession to Austin's contributions. Quine too, like Feyerabend and Wittgenstein, saw pursuit of nonsense as important to the progress of science, and central to the task of philosophy.

Much of his philosophy, properly understood, was involved in this pursuit. While it would take me too far afield to attempt to argue the point here, I shall simply state that this is how I view his mis-nomered "doctrine" of the indeterminacy of translation, as well as his talk of "posits".

Yet Austin was relatively uninteresting and unimportant to Quine, as he was to Wittgenstein, though for different reasons. For Quine Austin's brand of "lexicography", though a genuine contribution, was not all that central. Austin, he said, was, ironically, overattentive to a demarcation of disciplines, and left formal logic too far to the side of his considerations.[91] This, for Quine, arose from something more basic: Austin's "basic impatience with philosophical perplexity".[92]

This "impatience" was what disturbed Feyerabend as well. Feyerabend reads Austin as having pressed a certain "literalness" upon words to expose nonsense. This missed something Feyerabend felt he could find in Nestroy and Kraus, namely, an unmasking of "inhumanity". In this I feel that Feyerabend was responding to something special and unique about Austin, something upon which I would like to focus in closing.

What is to me most important about Austin lies, not in the "scientific" aspect of what he did, in his taxonomies and contrasts in and of themselves, but in the kind of self-understanding and self-confrontation that Austin's distinctions and examples convey and force upon the student of philosophy who reads him. He shows one, over and over again, that the way one ordinarily uses "I know" will not capture, express, or do justice to the epistemological question one was trying to ask. Nor will one's excuses, studied in their natural habitat of particular usage, involve one in any discussion of an intrinsically good will, or morality as such. He thereby tempts one to wonder whether there is anything at all that one is thinking of when one philosophizes about such things. (That there is nothing in it Austin suggests, but never claims to prove.) In no way can appeal to ordinary usage be the Last Word. But, as Austin said, it is not bad to notice it as the First Word. What is remarkable is that such noticing and describing is enough to accomplish what Austin accomplished. This allows his writings to raise the useful question whether there really is any more that ever can be accomplished in pure philosophy than to shift the burden on to the philosopher who wishes to resist him and depart from usage. Austin's method is the most pure form of anti-method in philosophy.

For Wittgenstein, Austin did not discuss (as Wittgenstein explicitly wrote to Moore) any "important points".[93] Austin, for his part, was known to have been impatient with Wittgenstein's remark in the *Investi-*

gations (§23) about there being a "countless" variety of uses of language. – So why not begin counting them?, one might well ask (unless one were to remember that Wittgenstein mentions here, as a "rough guide", the evolution of mathematics, and more or less evidently has in mind those parts in which counting, in Austin's sense, cannot, grammatically speaking, be accomplished).

If Austin and Wittgenstein are understood to share a philosophical method or subject matter, one had better leave room for substantial differences between them, differences of punctuation, emphasis and focus that are of distinct philosophical relevance, even and perhaps especially because the differences do not qualify as differences of opinion, or of grammatical analysis, or of subject matter, or of philosophical method or doctrine. Even when he made remarks that, taken in isolation, echo remarks of Austin's (as in some of *On Certainty*'s remarks about our uses of the verb "to know"), just about every so-called "grammatical" remark that Wittgenstein ever made is, next to Austin, wrong or coarse or hopelessly vague. (Besides, nearly every such remark Wittgenstein himself rounds on, in order to show that and how it is or may be made to seem ultimately irrelevant to the philosopher's concerns.) Austin's spirit has therefore naturally appeared to at least some of his readers (as it did to Quine and to Feyerabend) to be ultimately more like Frege's than like Wittgenstein's: Frege wrote to Wittgenstein that, unlike the author of the *Tractatus*, he had little interest in tracing those "disturbances of psychologico-linguistic origin" which lead us astray in speaking about the logic of language.[94] If a view such as Idealism is not true, then, for Frege, there can be no grounds for it all, hence it is a puzzle why anyone should be exploring it. Austin, in his unforgiving moments (and there are many of these), was given to the same form of impatience. He was also liable to credit as advances the contributions of logicians such as Carnap, instead of warning, as did Wittgenstein, against the dangers and seductiveness of their contributions.

It seems to me intrinsic to Austin's importance as a thinker that his philosophy, more than any other in history, defies encapsulation, as opposed to exemplification. By this I mean that Austin's philosophical power and acuity and contributions, observable in the particular cases with which he works, utterly vanish, in a way I think Wittgenstein's and Quine's do not, when one tries to explain, in a nutshell, why they are significant. This is why Austin himself offers us so few metaphors, analogies and suggestions about the nature of philosophy as such. To attempt to sum up the nature of his philosophical achievements by discussing his methods or subject matter, his historical place, his analogies, his observations, strategies or categorizations of speech acts – as opposed to applying them to concrete cases – is to lose sight of his achievements. What he seems to have left behind is practically nothing but the sense of an open field of investigation before us. I consider this to be a unique and significant philosophical achievement.

Thus, looking back at Feyerabend's use of Nestroy (and Kraus) against Austin, it seems that just as Feyerabend made a mistake in accusing Quine of a naïve general view of "science", he made a parallel mistake in having assumed that Austin depended in his philosophical work upon a general conception of "ordinary language" and/or "common sense". Feyerabend depicts Austin as evasive, refusing to answer legitimate objections to common sense philosophy ("There is not one common sense, there are many (I argued this point with Austin but could not convince him)."). But I imagine that Austin would have preferred Feyerabend to engage in a detailed, contextually-sensitive investigation of particular uses of the notion of "common sense". He would have wanted it made clear to Feyerabend that he (we) have no general notion of the "ordinary" or "commonsensical" in advance of this. This notion functions in a very complex way, is not an ordinary sortal like "shoe" or "hand", and we ought to look carefully at it, test it, and probe its appeal in particular circumstances, before we rush to philosophical judgment. Austin, just like Quine, would have refused to take a stand on Feyerabend's "different ways of knowing", scientific and non-scientific. He was too acute a practitioner of anti-method to enshrine any such large-scale distinction.

Austin felt that the next generation would be better served by carrying out a rather new and different project, an investigation of what we say when. When he counted and classified and recounted variations of usage, he hoped for more than the development of a syntax or grammar of a new and different kind. He hoped that philosophy teachers would teach something different, and differently, from much of what they had taught before. Austin was in this way, behind his British veneer of calm politeness, much more radical than Feyerabend, and far more radical than Quine or Wittgenstein. He really believed, and more or less said, that most philosophers were just making mistakes and wasting students' time. Unlike Wittgenstein, and something like Quine, he felt this without nostalgia or ambivalence about the past. Unlike Quine and unlike Wittgenstein, Austin bothered to try to illustrate and document the uninterestingness of traditional epistemological questions in graphic detail.

What makes Wittgenstein *seem* more deep than Austin, and what makes him in fact more like Feyerabend, is his explicit acknowledgement that the grammatical observations one makes are drenched, always, with value, context, and matters of individual idiosyncracy. I believe that Wittgenstein came to appreciate this difference between himself and Austin; the following passage in his later remarks on psychology reeks of a sharp difference with Austin:

> What is it that is repulsive in the idea that we study the use of a word, point to mistakes in the description of this use and so on? First and foremost one asks oneself: How could *that* be so important to us? It depends on whether what one calls a 'wrong description' is a description that does not accord with established usage – or one which does not accord with the practice of the person giving the description. Only in the second case does a philosophical conflict arise.[95]

A philosophical problem, for Wittgenstein, has an essentially first-person form, viz., "I don't know my way about" or "I don't recognize myself" – in my concepts, in my problems, in my language, in my questions. Wittgenstein saw that Austin's cataloguing of ordinary usage stopped in a place that philosophy could not comfortably stand. If you like, Wittgenstein, like Feyerabend, had a greater awareness of the central importance to philosophy, not only of the contextual specificity of grammar, but also of the value-ladenness of remarks about the limits of sense. Wittgenstein attempted to do justice to the need people feel to go back to the sources of, and need for, general questions, precisely *because* there is in the end nothing general to be said that cannot also mislead.

Progress in philosophy can look greater than it is. But, as I hope to have at least suggested in this essay, it can be greater, and more various, than it sometimes looks.

Notes

1. I am grateful to Stanley Cavell, Robert S. Cohen, Jeff Coulter, Kurt Fischer, Allan Janik, Akihiro Kanamori, Elisabeth Nemeth, Hilary Putnam, and Friedrich Stadler for conversations about Feyerabend; this essay could not have been written, in particular, without Fischer's having afforded me the opportunity of putting my ideas before

his seminar at the University of Vienna on "Austrian Philosophy" and Stadler's invitation to present a version of this paper at the Institut Wiener Kreis conference on Feyerabend in June 2004. I would also like to thank the audience at the Mind and Society Group in Cambridge, England, August 2004, for their feedback on yet another version of the paper, and Roger Gibson for discussion of Quine. A late draft of the paper benefited from suggestions of Robert Bowditch, Nadine Cipa, Eric Oberheim, Rohit Parikh, and David Stern.

My comparative approach, especially as applied to Quine and Austin, owes much to many years of conversation with Burton Dreben. In particular I am indebted to his May 1998 lecture to the Canadian Philosophical Association, "Austin, Quine, Wittgenstein: Are They Doing Philosophy?", a performance resulting from seminars he gave at the University of Oslo and Boston University. This essay is in many ways my reply to Burt.

This research was generously supported with the help of Megan Burns, Keren Gorodetzky, a sabbatical leave from Boston University and fellowships awarded to me by the American Philosophical Society, the Wellesley College Stevens Fellowship, and the Fulbright Foundation.

2. I have articulated some of these larger themes elsewhere, especially in the Introduction (with Sanford Shieh) to *Future Pasts: Perspectives on the Place of the Analytic Tradition in Twentieth Century Philosophy* (New York: Oxford University Press, 2001), pp. 3-21 and in "The Fact of Judgment: The Kantian Response to the Humean Condition", in J. Malpas, ed., *From Kant to Davidson: Philosophy and the Idea of the Transcendental* (London: Routledge, 2003), pp. 22-47 and "'Putnam's "The Meaning of 'Meaning'": Externalism In Historical Context", in *Philosophers in Focus: Hilary Putnam*, ed. Y. Ben Menachem (Cambridge University Press), pp. 17-52.

3. Cf. Kurt Rudolf Fischer, "Paul Feyerabend: A Personal Reminiscence", in *Aufsätze zur angloamerikanischen und österreichischen Philosophie* (Peter Lang, 1999), pp. 77-90, especially p. 87: "Wittgenstein, I believe, was the only philosopher whom Feyerabend never or almost never criticized."

4. See, for example, the following remark in Feyerabend's "Concluding Unphilosophical Conversation" in Gonzalo Munévar, ed., *Beyond Reason: Essays on the Philosophy of Paul Feyerabend, Boston Studies in the Philosophy of Science vol. 132* (Dordrecht: Kluwer Academic Publishers, 1991), pp. 489-528, specifically p. 489:

B: [*Against Method* says that it is] a letter, a personal communication; not a treatise, not a textbook. A letter written tongue in cheek.
A: You mean the whole book is a joke?
B: No. I am serious about lots of things in it; but summing them up in the form of a philosophical "position" – that was the joke. Many reviewers fell for it – although I left a sufficient number of clues ...
A: Now, wait a minute! You say you have some serious points to make.
B: Yes.
A: But you don't have a philosophical position.
B: No. I may have had something like a philosophical position when I was a student, and early in my career. At that time I thought there was no knowledge but scientific knowledge and all the rest was bunk. That is kind of a "position", isn't it?
A: And then you became an anarchist.
B: No. Then I read Wittgenstein.
A: Wittgenstein?

B: Yes. I read his *Remarks on the Foundation* [sic] *of Mathematics* and his *Philosophical Investigations* in manuscript, various versions of them, years before they appeared in print and I discussed the content with Elizabeth Anscombe who was then in Vienna, to learn German for her translation of Wittgenstein's works. Incidentally, I studied Wittgenstein's writings much more thoroughly than anything from the Popperian inventory, though there are still some people who regard me as a Popperian apostate.

5. Wittgenstein, preface to *Philosophical Remarks* (Chicago: University of Chicago Press, 1975).
6. Cf. Wittgenstein, *On Certainty* (eds. G.E.M. Anscombe, G.H. von Wright, New York: Harper Torchbooks, 1969) §422. For Feyerabend on pragmatism, we shall consider his remarks about Quine below.
7. Although most readers have taken Wittgenstein's *Tractatus* to forward a version of these latter errors, rather than an attempted critique of them, like Feyerabend I take the *Tractatus* to have been aiming, however imperfectly, at this kind of critique. On Feyerabend's views of the *Tractatus*, see section 2. below. For some of mine, see "The Uncaptive Eye: Solipsism in the *Tractatus*" in L. Rouner, ed., *Loneliness* (Notre Dame: Boston Studies in the Philosophy of Religion, 1998), pp. 79-108, "Number and Ascriptions of Number in Wittgenstein's *Tractatus*", in Floyd and Shieh eds., *Future Pasts: Perspectives on the Analytic Tradition in Twentieth Century Philosophy*, pp. 145-192 and "Wittgenstein on Philosophy of Logic and Mathematics" in S. Shapiro, ed., *Oxford Handbook to the Philosophy of Logic and Mathematics*, (Oxford University Press, 2005), pp. 75-128.
8. This is a major theme in Wittgenstein's *Remarks on the Foundations of Mathematics*. Feyerabend makes the point explicitly in his "Concluding Unphilosophical Conversation", in Gonzalo Munévar, ed., *Beyond Reason: Essays on the Philosophy of Paul Feyerabend, Boston Studies in the Philosophy of Science vol. 132* (Dordrecht: Kluwer Academic Publishers, 1991), pp. 489-528; cf. especially p. 514.
9. An especially valuable essay for setting Feyerabend into context against recent discussions of post-modern philosophy of science – an essay that has the virtue of taking into account the full panoply of Feyerabend's writings, published and unpublished – is John Preston's "Science as Supermarket: 'Post-Modern' Themes in Paul Feyerabend's Later Philosophy of Science", in *The Worst Enemy of Science? Essays in Memory of Paul Feyerabend*, eds. J. Preston, G. Munevar, D. Lamb (New York: Oxford University Press, 2000), pp. 80-101.
10. I am assuming readers of Austin's "Other Minds" will recognize this as more or less obviously shaping his discussions of knowing (see Austin, *Philosophical Papers*, J.O. Urmson and G.J. Warnock, eds. (3rd. ed., New York: Oxford University Press, 1979), pp. 76-116). Less appreciated may be that this is also the point (though not the whole context) of Quine's remark ("Relativism and Absolutism", *The Monist* 67 (1984): 293-296, p. 295) that "'Know' is like 'big': useful and unobjectionable in the vernacular where we acquiesce in vagueness, but unsuited to technical use because lacking in a precise boundary. Epistemology, or the theory of knowledge, blushes for its name."
11. Compare Quine's *Pursuit of Truth* (2nd ed. Harvard University Press, 1992), §1: "Not that prediction is the main purpose of science. One major purpose is understanding."
12. There are those, of course, who would argue that Austin and Quine are also "end of philosophy" or purely negative thinkers. I am not going to discuss this issue here, though it is a fascinating one.
13. Compare Peter Hacker, who has recently claimed that, at its best, analytic philosophy should be essentially prosecutorial, conceived as "*sui generis*, as a critical disci-

pline *toto caelo* distinct from science, as an *a priori* investigation, as a tribunal of sense as opposed to a plaintiff confronting nature," "before which science should be arraigned when it slides into myth-making and sinks into conceptual confusion." (P.M.S. Hacker, "Analytic Philosophy: What, Whence, and Whither?" in A. Biletzki and A. Matar, eds., *The Story of Analytic Philosophy: Plots and Heroes* (London: Routledge, 1998), pp. 3-36, see especially pp. 25 and 29.) Compare Hacker's *Wittgenstein's Place in Twentieth-Century Analytic Philosophy* (Oxford and Cambridge, Mass.: Blackwell, 1996), especially his attack on Quine. Sanford Shieh and I offer a contrasting perspective on the analytic tradition in our introduction to *Future Pasts: Perspectives on the Place of the Analytic Tradition in Twentieth Century Philosophy*, pp. 3-21.
14. From "Our Method", in G. Baker ed., G. Baker et al. trans., *The Voices of Wittgenstein: The Vienna Circle* (New York: Routledge, 2003), p. 277.
15. Paul K. Feyerabend, *Three Dialogues on Knowledge* (New York: Oxford University Press, 1991), pp. 49ff. Compare Fischer's discussion in his "Paul K. Feyerabend: A Personal Reminiscence", pp. 79-80.
16. I have written more about this theme in my introduction with Sanford Shieh to *Future Pasts: The Analytic Tradition in Twentieth Century Philosophy*. On Quine's "extensionalism" about people, see his "Confessions of a Confirmed Extensionalist", in Floyd and Shieh, eds., *Future Pasts*, pp. 215-222. The reader might compare Rawls's "Afterward" to this book for the expression of some similar ideas about how to read historical texts.
17. In the edition of *Against Method* that I first read (1st ed., London: Verso, 1972), Wittgenstein is mentioned only once, in a context where psychology and history are brought in as correctives to a view of evidence Feyerabend attributes to him, a view invested with a strong "incommensurability" aspect. See p. 133: "... Such disregard for phenomena, which for us are quite obvious, may be due either to a certain indifference towards the existing evidence, which was, however, as clear and as detailed as it is today, *or else to a difference in the evidence itself*. It is not easy to choose between these alternatives. Having been influenced by Wittgenstein, Hanson, and others, I was for some time inclined towards the second version, but it now seems to me that it is ruled out both by physiology (psychology) ... and by historical information." For an interesting criticism of this remark that is similar to criticisms of Feyerabend made by Putnam in the mid-1960s (see section 3. below), see Jeff Coulter, *Rethinking Cognitive Theory* (New York: St. Martin's Press, 1983), p. 126, n2.
18. Fischer, "Paul K. Feyerabend: A Personal Reminiscence", pp. 85-86.
19. Feyerabend was open to alternative medicinal approaches as much out of practical need as out of intellectual curiosity. See Fischer, p. 86. Robert Cohen has told me that on Feyerabend's visits to Boston, he would often ask to have Chef Joyce Chen (a local luminary and close personal friend of Cohen's) prepare him dishes that were especially easy on his often irritated stomach.
20. See Paul K. Feyerabend, "Last Interview", by Joachim Jung, in *The Worst Enemy of Science? Essays in Memory of Paul Feyerabend*, eds. J. Preston, G. Munevar, D. Lamb (op.cit.), pp. 159-168 (reference is to p. 164).
21. See the Preface to *Against Method*.
22. In "Science as Supermarket: 'Post-Modern' Themes in Paul Feyerabend's Later Philosophy of Science" (op.cit.) Preston explicitly raises the issue of Feyerabend's disingenuousness in presenting a rewriting of the history of his thinking about meaning (p. 87). At the same time, Preston is concerned to articulate and defend a sympathetic portrait of Feyerabend's significance as a philosopher of science. It is noteworthy that, by emphasizing Feyerabend's focus on the "disunity" of science, Pre-

ston ends by calling Feyerabend a "Wittgensteinian" (p. 99). Compare my discussion of Feyerabend on Quine and Austin in section 4 below.
23. See Paul Feyerabend, *Killing Time: The Autobiography of Paul Feyerabend* (Chicago: University of Chicago Press, 1995), pp. 30, 108, 119.
24. *Killing Time*, pp., 64-67.
25. See *Killing Time*, pp. 68ff. Compare Preston's discussion of Feyerabend's later insistence on science as art in "Science as Supermarket: 'Postmodern' Themes in Paul Feyerabend's Later Philosophy of Science", especially at pp. 94ff.
26. See Feyerabend's added note to his review of *Philosophical Investigations*, from 1980, where he points out this connection explicitly. Rupert Read emphasizes the evolution of this Feyerabend/Kuhn contrast in his "On Wanting to Say, 'All We Need is a Paradigm'", *Harvard Review of Philosophy* IX (2001): 88-105. In particular, Read offers an instructive interpretation of Feyerabend's pro-Kuhnian remarks in two interesting later works, viz., "Two Letters of Paul Feyerabend to Thomas S. Kuhn on Draft of *Structure*", Paul Hoyningen-Huene, ed., *Studies in the History and Philosophy of Science* 26, 3 (1995): 354-5 and "More Clothes from the Emperor's Bargain Basement; A Review of Larry Laudan's 'Progress and its Problems,'" in *Problems of Empiricism Collected Papers Volume 2* (New York: Cambridge University Press, 1981) pp. 236-7.
27. *Killing Time*, pp. 152, 179.
28. Compare Ian Hacking, Obituary of Paul Feyerabend, *Common Knowledge* 3, 2 (1994): 23-28.
29. Compare Preston, "Science as Supermarket: 'Post-Modern' Themes in Paul Feyerabend's Later Philosophy of Science" (op.cit.) for a discussion of what Preston calls Feyerabend's "semantic nihilism". As Putnam pointed out in the mid-1960s (see section 3. below), Preston notes that Feyerabend sometimes held theories of meaning at the same time as he objected to the very idea of a theory of meaning.
30. Compare Fischer, "Paul K. Feyerabend: A Personal Reminiscence", p. 89: "Feyerabend sees the attack on Essentialism as historically relative, confined to our time. And this is the only point I know of at which Feyerabend criticizes Wittgenstein. Feyerabend writes ... 'Finally, a critical comment on Wittgenstein's idea of philosophy. Wittgenstein assumes that philosophers want to provide a theory of already existing things, and he is correct in pointing out that what exists is much more complicated than any philosophical theory. However, philosophical theories have not merely reflected things but have changed them, i.e. the (sham) conflict between theory and practice was resolved by a change of practice. This fact refutes the idea that philosophers, and for that matter all mythmakers, only erect castles in the air, and introduces a fruitful relativism of the kind explained in my *Erkenntnis für freie Menschen* (Frankfurt, 1980).'" (added note to Feyerabend's "Review of Wittgenstein's *Philosophical Investigations*", reprinted in *Problems of Empiricism*, pp. 99-130; quote from p. 130).
31. Indeed, it was these remarks which appear to have initially drawn Feyerabend to Wittgenstein; cf. footnote 4.
32. Kant, *Metaphysical Foundations of Natural Science*, Vol. IV of *Kants Gesammelte Schriften* (ed. Royal German Academy of Sciences, Berlin, Walter deGruyter & Co., 1900-), pp. 470ff.
33. See Wittgenstein, "Review of P. Coffey, *The Science of Logic*", reprinted in J. Klagge and A. Nordmann eds., *Philosophical Occasions 1912–1951* (Indianapolis: Hackett, 1993), pp. 1-2.

34. Compare the discussion of the particular case of Gödel's theorem in my "Prose versus Proof: Wittgenstein on Gödel, Tarski and Truth", *Philosophia Mathematica* (3) vol. 9 (2001): 901-928.
35. Compare Peter Hylton, *Russell, Idealism and the Emergence of Analytic Philosophy* (Oxford, The Clarendon Press, 1990).
36. I have in mind here Michael Dummett, *Origins of Analytical Philosophy* (Cambridge, MA: Harvard University Press, 1993), Richard Rorty, *Philosophy and the Mirror of Nature* (Princeton, N.J.: Princeton University Press, 1979) and Scott Soames, *Philosophical Analysis in the Twentieth Century* (2 vols., Princeton, N.J.: Princeton University Press, 2003).
37. Fischer, "Paul K. Feyerabend: A Personal Reminiscence", p. 80.
38. Fischer, "Paul K. Feyerabend: A Personal Reminiscence", p. 80.
39. See *Tractatus* Preface and 4.002 and compare Paul Ernst, „Nachwort", in P. Ernst ed., *Kinder- und Hausmärchen gesammelt durch die Brüder Grimm (Die Grimmschen Märchen), Band 3* (München und Leipzig, Georg Müller, 1900): pp. 271-314, especially pp. 307-308. Wittgenstein wrote (*Wiener Ausgabe* (New York: Springer Verlag, 1993–), ed. M. Nedo, Vol. III, p. 266): "If my book is ever published, it must contain in its Preface an allusion to Paul Ernst's Preface to the Grimm's Fairy Tales, which I ought to have mentioned in the Log. Phil. Abhandlung as the source of the expression, 'misunderstanding of the logic of language.'"
40. *Killing Time*, p. 103.
41. *Killing Time*, p. 63.
42. An important passage for grasping the ethical flavor of the rule-following passages in the later philosophy is *Remarks on the Foundations of Mathematics* (G.H. von Wright, R. Rhees, G.E.M. Anscombe, eds., G.E.M. Anscombe, trans., revised edition, Cambridge, MA: 1978) I §13: "It strikes us as if something else, something over and above the *use* of the word 'all', must have changed if "*fa*" is no longer to follow from "$(x).fx$"; something attaching to the word itself. Isn't this like saying: 'If this man were to act differently, his character would have to be different'. Now this may mean something in some cases and not in others. We say 'behaviour flows from character' and that is how use flows from meaning." A good man can act on occasion badly, and a bad man well – but not beyond a certain point, a point we can determine only by investigating particular examples.
43. *Killing Time*, p. 75.
44. Some of this has been documented in a book about Wittgenstein's sister, *Margaret Stonborough-Wittgenstein*, by Ursula Prokop (Böhlau, 2003); cf. Ray Monk, *Ludwig Wittgenstein: The Duty of Genius* (New York: Free Press, 1990), pp. 562ff.
45. The chapter of *Killing Time* on Feyerabend's political experiences up to and during the war is, by contrast, correspondingly flat and uninteresting, despite his effort to be unsparing and unromantic about his youth (compare Fischer, "Paul K. Feyerabend: A Reminiscence", p. 82). It ends up being anti-dramatic, as if, there having been nothing in particular he felt he could say, he decided to stay silent, and let others draw their own conclusions. It should be compared with a passage in Feyerabend's "Concluding Unphilosophical Conversation", p. 511:

    A: Do you mean to say that you are opposed to condemning the atrocities of the Nazi era?
    B: I am – if the condemnation is pronounced in empty space, as it were, on the basis of superficial and loaded facts and if it is demanded of people who have no emotional contact with the events and the victims. A "moral condemnation" of this kind is a meaningless curse, the request to repeat it an imposition, and

any action undertaken on its basis a crime. Many so-called educators in present day Germany don't seem to realize this.
A: It is an empty curse to condemn Auschwitz?
B: If the words have no connection with personal experiences, fears, expectations – yes. The past cannot be conquered and should not be judged except by those who are willing to enter it.
A: But that is impossible ...
B: For a philosopher, or an "objective" historian, yes. But a poet, a novelist, a filmmaker, given suitable material can recreate the atmosphere; he or she can bring to life the terror, the cruelty as well as the fascination of the time and thus lay the ground for a genuine moral decision.

46. *Killing Time*, p. 76. Ray Monk has described this as "Probably the only public meeting of philosophers that Wittgenstein attended while he was in Vienna" during this final visit (*The Duty of Genius*, p. 563).
47. *Killing Time*, p. 93.
48. Cavell's "Ending the Waiting Game: A Reading of Beckett's *Endgame*", written in 1964, is the *locus classicus* here; it may be found as chapter V of his *Must We Mean What We Say?* (New York: Cambridge University Press, updated edition 2002). The topics of modernism, Wittgenstein and Beckett are touched on elsewhere in this book of essays, most obviously in "A Matter of Meaning It" (1965) (cf. p. 219) and "The Avoidance of Love: A Reading of *King Lear*" (1966-67) (cf. p. 349). The opening of the *Investigations* is specifically treated along with Beckett in Cavell's "Notes and Afterthoughts on the Opening of Wittgenstein's *Investigations*", an updated version of lectures notes begun at Berkeley in 1960, in *Philosophical Passages: Wittgenstein, Emerson, Austin, Derrida* (Cambridge, MA: Basil Blackwell, 1995), pp. 125-186 (cf. especially pp. 169-70).
49. *Killing Time*, p. 113.
50. Compare Cavell, "Ending the Waiting Game", pp. 117ff.
51. Compare Cavell, "Ending the Waiting Game", p. 120 and *The Claim of Reason: Wittgenstein, Skepticism, Morality, and Tragedy* (Oxford University Press, 1979) *passim*, on the theme of Wittgenstein on privacy.
52. Wittgenstein, by contrast, wrote explicitly about his troubles understanding Shakespeare (see *Culture and Value*, trans. P. Winch (Oxford: Blackwells, revised 2nd edition, 1998). Oddly enough, he too seems to have had trouble conceiving of Shakespeare as great, explicitly complaining at one point about Shakespeare's lack of realism. Perhaps, as David Stern suggested to me in conversation, this particular block was another Austrian thing.
53. Cavell links Wittgenstein, via these literary forms, to the romantic tradition, and specifically to figures like Schlegel; see Cavell's "Declining Decline: Wittgenstein as a Philosopher of Culture" (written 1986) in *This New Yet Unapproachable America: Lectures after Emerson after Wittgenstein* (Albuquerque, NM: Living Batch Press, 1989). So far as I know none of these post-Kantian Idealists was ever mentioned by Feyerabend.
54. *Killing Time*, p. 93. The original essay, translated into English by G.E.M. Anscombe, was written in 1952 and published in *The Philosophical Review* 64, 3 (July 1955): 449-483. Feyerabend added some remarks to the essay when it was reprinted in his *Problems of Empiricism: Collected Papers Volume 2*, pp. 99-130. Page references will be to this later edition of this work.
55. George Pitcher, ed., *Wittgenstein: The Philosophical Investigations* (Garden City, NY: Anchor Books, Doubleday, 1966).

56. Feyerabend, "Review of Wittgenstein's *Philosophical Investigations*", p. 126. Paul Churchland points out that Feyerabend's later epistemological emphasis on a pluralism of approaches (epistemological anarchism) fits well with the practice of cognitive scientists who appeal to the complexity of neural networks to understand the inevitably theoretically saturated nature of perception ("To Transform the Phenomena: Feyerabend, Proliferation, and Recurrent Neural Networks", in *The Worst Enemy of Science?*, eds. Preston, Munévar and Lamb (op.cit.), pp. 148-158). However, we ought to note that in such a model considerations of language and its modules would still play a more complicated role than Feyerabend might have let on. In the early review of Wittgenstein, it seems to me that we might at best see Feyerabend groping toward a kind of Quinean anti-reductionist monism about mental states of precisely the kind Churchland defends elsewhere – but crucially lacking Quine's, Churchland's (and Wittgenstein's) sophisticated concerns about the extent to which concepts such as meaning and truth can be scientifically explicated.

   Feyerabend's insight into Wittgenstein's attack on the whole idea of understanding as a mental state may be compared with Putnam's insight that the slogan of his 1975 essay on externalism, "meanings ain't in the head", is a slogan that, despite its uses, can mislead in inviting the (misguided) question, "well then where are these entities, if not in the head?". For more on this topic, see my "Putnam's 'The Meaning of "Meaning"': Externalism In Historical Context" (op.cit.), especially pp. 23ff. For an excellent essay on why Wittgenstein's attack on the idea of understanding as a mental state is anti-scientistic, without being anti-scientific, see Warren Goldfarb, in *Midwest Studies in Philosophy XVII: The Wittgenstein Legacy*, P.A. French, T. E. Uehling, Jr. and H.K. Wettstein (eds.): 109-122.
57. The reader may compare Anscombe's *An Introduction to Wittgenstein's Tractatus* (Philadelphia, PA: University of Pennsylvania Press, 1st ed. 1959, 3rd ed. 1971); especially (in the 3rd edition) pp. 12-13.
58. Feyerabend, "Review of Wittgenstein's *Philosophical Investigations*", p. 128.
59. On this see Cavell's "Aesthetic Problems of Modern Philosophy", begun in 1962, in *Must We Mean What We Say?*, pp. 73-96. I give a reading of Kant and Wittgenstein informed by Cavell's interpretive suggestions in my "Heautonomy and the Critique of Sound Judgment: Kant on Reflective Judgment and Systematicity", *Kants Ästhetik / Kant's Aesthetics / L'Esthétique de Kant*, Herman Parret, Hrsg./ed. (Berlin–New York, Walter de Gruyter Verlag, 1998), pp. 192-218.
60. Feyerabend, "Review of Wittgenstein's *Philosophical Investigations*", p. 123.
61. Feyerabend, "Review of Wittgenstein's *Philosophical Investigations*", p. 124.
62. Feyerabend, "Review of Wittgenstein's *Philosophical Investigations*", pp. 111n, 126, 127.
63. Jaakko Hintikka has gone so far as to remark that the idea of logic as analytic and empty of empirical content, which made Wittgenstein famous, was a very widespread, indeed hackneyed idea among Austrians at the turn of the century; he appeals to Mach as his proof text. See Hintikka, "Ernst Mach at the Crossroads of Twentieth-Century Philosophy", in Floyd and Shieh, eds., *Future Pasts: The Analytic Tradition in Twentieth Century Philosophy*, pp. 81-100.
64. "A Note on the Paradox of Analysis", in *Philosophical Studies* VII, 6 (1956): 92-96.
65. For a lucid discussion of these early uses and abuses of Wittgenstein by Feyerabend, see John Preston, *Feyerabend: Philosophy, Science and Society* (Malden, MA: Blackwell, 1997), especially chapter 2.
66. Quine, "Progress on Two Fronts", reprinted in *Quintessence: Basic Readings From the Philosophy of W.V. Quine* (Harvard University Press, 2004), pp. 169-176. See also §3 of Quine's *Pursuit of Truth* (op.cit.). The terminology of "holophrastic" and

"analytic" derives, interestingly, from C.I. Lewis's *An Analysis of Knowledge and Valuation*, but it is against Lewis's notion of the hardness of the "given" and the conventionality of "meaning" that Quine is directing his uses of the distinction (and much else in his philosophy besides).
67. Quine, "Progress on Two Fronts", p. 173. See also his remarks on Kuhn, Hanson and Feyerabend in "I, You, and It: An Epistemological Triangle", in A. Orenstein and P. Kotatko eds., *Knowledge, Language and Logic* (Dordrecht: Kluwer Academic Publishers, 2000), pp. 1-6, esp. p. 5.
68. Quine's "Relativism and Absolutism" *The Monist* 67 (1984): 293-296 attempts to address it.
69. See Quine's "Five Milestones of Empiricism", in *Theories and Things* (Cambridge, MA: Harvard University Press, 1981), pp. 67-72; quotation from p. 67.
70. There is, unfortunately, relatively little written on the useful interplay and overlap between Quine and Wittgenstein. Readers more often than not either dismiss them altogether jointly, as "meaning skeptics" (which they were not), or choose one over the other as a hero. The essays by Burton Dreben, Roger F. Gibson and Douglas Winblad in *Wittgenstein and Quine*, eds. R. L. Arrington and J.-J. Glock (New York: Routledge, 1996) are exceptions.
71. The criticisms of Feyerabend are articulated in Putnam's, "How Not to Talk about Meaning", first published in R. Cohen and M. Wartofsky eds., *Boston Studies in the Philosophy of Science*, vol. 2: *In Honor of Philipp Frank* (New York: Humanities Press Inc., 1965) and reprinted in Putnam's *Mind, Language and Reality: Philosophical Papers Vol. 2* (New York: Cambridge University Press, 1975), pp. 117-131. My page references are to the latter edition.
72. I discuss this anti-rationalist aspect of externalism in my "Putnam's "The Meaning of 'Meaning'": Externalism In Historical Context", op.cit.
73. Putnam, "How Not to Talk About Meaning", pp. 125-6.
74. Compare the anecdote told by Putnam about Hempel, who spoke of a formalism "inheriting the disease" of ordinary language, in J. Floyd and H. Putnam, "A Note on Wittgenstein's 'Notorious Paragraph' about the Gödel Theorem", *Journal of Philosophy* 45, 11 (2000): 624-632.
75. *Philosophical Investigations* §§123-124, quoted in full, run thus:
"123. A philosophical problem has the form: 'I don't know my way about'.
124. Philosophy may in no way interfere with the actual use of language; it can in the end only describe it. For it cannot give it any foundation either. It leaves everything as it is. It also leaves mathematics as it is, and no mathematical discovery can advance it. A 'leading problem of mathematical logic' is for us a problem of mathematics like any other."
"A leading problem" alludes specifically to Ramsey's paper solving a portion of the decision problem ("On A Problem of Formal Logic" (1928), in F. P. Ramsey, *The Foundations of Mathematics* (London: Routledge and Kegan Paul, 1931, ed. R.B. Braithwaite, pp. 82-111)). I take Wittgenstein to be partly directing his criticism at Ramsey, but also partly at the *Tractatus*, which had unwittingly appeared, as Ramsey more than likely suggested to Wittgenstein, to hinge an account of the nature of logic on a positive outcome of the *Entscheidungsproblem*.
76. It also echoes and problematizes, somewhat ironically, both the *Tractatus* remark about how little has been done when philosophical problems are resolved and its claim, in the same breath, to have provided unassailably true solutions to those problems.
77. That talk about everything, about all objects at once, as such, faces more or less evident difficulties was a theme of Wittgenstein's (and not just Wittgenstein's) phi-

losophy from the beginning. So this remark ought to be compared, for example, with the opening lines of the *Tractatus*. It resonates too with Quine's joke about the ontological question, framed at the opening of his essay "On What There Is" (in *From a Logical Point of View* (2nd edition, Cambridge, Ma: Harvard University Press, 1980), pp. 1-19), p. 1: "A curious thing about the ontological problem is its simplicity. It can be put in three Anglo-Saxon monosyllables: 'What is there?' It can be answered, moreover, in a word – 'Everything' – and everyone will accept this answer as true. However, this is merely to say that there is what there is. There remains room for disagreement over cases; and so the issue has stayed alive down the centuries." As Quine well knew, "Everything" is not a sentence, true or false, but a word. His remark, like *Investigations* §124, asks us to ponder the conditions and assumptions informing talk about "everything".

78. David G. Stern has rightly stressed how often readers of Wittgenstein have rested with a far too simplistic understanding of the motto to the *Investigations*. Stern takes a far more sophisticated and satisfying reading of the motto as a point of departure for his reading of the book in *Wittgenstein's Philosophical Investigations: An Introduction* (New York: Cambridge University Press, 2004).
79. *Killing Time*, pp. 143-144.
80. The full context of Quine's remark is in "Mr. Strawson on Logical Theory", in *The Ways of Paradox and Other Essays* (Cambridge, MA: Harvard University Press, revised ed. 1976), pp. 137-157, specifically pp. 150-151:
"The long and perceptive passages in which Mr. Strawson traces out something like a logic of ordinary language have all the interest and value of an able philological inquiry. But it is a mistake to think of Mr. Strawson as doing here, realistically, a job which the dream-beset formal logician had been trying to do in his unrealistic way. Actually the formal logician's job is very different ... He does not care how inadequate his logical notation is as a reflexion of the vernacular, as long as it can be made to serve all the particular needs for which he, in his scientific program, would have otherwise to depend on that part of the vernacular ...
Not that this logical language is independent of ordinary language. It has its roots in ordinary language, and these roots are not to be severed.
If certain problems of ontology, say, or modality or causality, or contrary-to-fact conditionals, which arise in ordinary language, turn out not to arise in science as reconstituted with the help of formal logic, then those philosophical problems have in an important sense been solved: they have been shown not to be implicated in any necessary foundation of science. Such solutions are good to just the extent that (a) philosophy of science is philosophy enough and (b) the refashioned logical underpinnings of science do not engender new philosophical problems of their own."
81. See, for example, *Word and Object* (Cambridge, MA: M.I.T. Press, 1960), chapter 1.
82. On this see especially Quine's "Has Philosophy Lost Contact With People", a reply to Mortimer Adler, reprinted in *Theories and Things*, pp. 190-193.
83. "Last Interview", p. 162.
84. An exception already mentioned is his brief and very early piece, "A Note on the Paradox of Analysis", which seems to me one of Feyerabend's weakest essays.
85. A perfect example of such recasting is offered in Quine's "Five Milestones of Empiricism", reprinted in *Theories and Things*, pp. 67-72.
86. Consider the distinction between analytic and synthetic, which Quine became famous undermining. It is important to remember that Quine was prepared to grant a certain sense to the difference between a merely verbal difference and a difference of substance, or a piece of language whose content is learned with the learning of the words, rather than with an investigation of reality. He simply did not wish to have

philosophers erect needlessly over-theorized accounts of them. Compare *The Roots of Reference: The Paul Carus Lectures* (La Salle, IL: Open Court, 1973) §21.
87. Wittgenstein (especially early Wittgenstein) may not be comprehensible, as Quine and Austin more obviously are, as a philosopher in love with the vernacular. Whether or not one might read the *Tractatus* and/or *Philosophical Investigations* as an effort to move everyday speech back into metaphysical discussion, or vice versa, is a large issue that I cannot survey here.
88. "On Austin's Method", in *Theories and Things*, pp. 86-91, quotation from p. 86.
89. Fischer, "Paul K. Feyerabend: A Personal Reminiscence", p. 80.
90. Quine, "On Austin's Method", pp. 87, 91.
91. Quine, "On Austin's Method", p. 89.
92. Quine, "On Austin's Method", p. 89.
93. See Wittgenstein's letter to Moore of 3 December 1946 (in Wittgenstein, *Cambridge Letters*, eds. B. McGuinness and G.H. von Wright (Cambridge, MA: Blackwells, 1995), p. 325) : "[...] Price at the last Mor[al] Sciences] Cl[ub] meeting was *by far* better than Austin had been. Price was willing to discuss important points."
94. See the letter from Frege to Wittgenstein of 3 April 1920, published on the CD-rom of Wittgenstein's *Briefwechsel*, eds. Monika Seekircher, Brian McGuinness and Anton Unterkircher (Intelex, Innsbrucker elektronische Ausgabe 2004). A translation into English of the letter by myself and B. Dreben is in the volume *Successor and Friend: Georg Henrik von Wright and Ludwig Wittgenstein*, ed. E. De Pellegrin (New York: Springer Verlag, forthcoming).
95. Wittgenstein, *Remarks on the Philosophy of Psychology Vol. I*, G.E.M. Anscombe and G.H. von Wright, eds., G.E.M. Anscombe, trans. (University of Chicago Press, 1980) §548.

NAMENREGISTER / INDEX OF NAMES

Nicht erfasst wurden Anmerkungen und Literaturverzeichnisse, Tabellen und Graphiken.
Notes, references, tables and figures are not included.

Acham, K. xxi
Adorno, T.W. xviii, xxi, 4
Agamben, G. 53
Albert, H. x, xvii, xxi, 7, 8
Anscombe, E. xi, xii, 15, 105, 117, 118, 121-123
Aristoteles (Aristotle) 35, 36, 78
Armitage, A. 78
Aspect, A. 88
Augustine 120
Austeda, F. xvii
Austin, J.L. 7, 99, 100, 101, 102, 103, 104, 110, 111, 112, 113, 127, 129, 130, 131, 132, 133, 134, 135, 136, 137, 138, 139, 140, 141
Ayer, A. 3, 131
Bachmann, I. 57
Bakunin, M. 64
Ball, H. 52
Beckett, S. 120
Bell, J.S. 84, 85, 88
Benedict, R. 66
Benedikt, M. 9
Bergson, H. 78
Berkeley, G. 9
Birkhoff, G. 77, 84
Bloch, E. xviii, xxi
Bohm, D. 81, 86
Bohr, N. xv, xxv, 19, 81-83, 85, 86, 105
Boltzmann, L. xxiv, 35, 36, 40, 46
Boole, G. 84
Borrini-Feyerabend, G. viii
Brecht, B. xix, 38, 115
Brezinova, I. 79
Bridgman, P.W. 88
Broad, C.D. 78
Broglie, L. de 87
Brunelleschi, F. 55
Brunswik, E. 1, 6
Büchner, L. 2
Canetti, E. 56
Carnap, R. x, xv, xviii, xxi, 125, 139
Cassirer, E. xv
Castaneda, C. 66
Cavell, S. 120, 123

Copernicus (Kopernikus), N. xx, 43, 78
Dalí, S. 77
Dedekind, R. 110
Demokrit 57
Dempf, A. x
Dennis, W. 16
Descartes, R. 2, 9, 72
Dijksterhuis, E.J. 35
Dingler, H. xi
Dirac, P. xviii, 88
Duhem, P. xi, xiv, 13, 14, 15, 16, 23, 46
Dummett, M. 99
Durkheim, E. 57
Dyson, F. 76
Ehrenhaft, F. x, 21, 108
Eichhorn, H. x
Einstein, A. xxv, xxvi, 80, 81, 83, 84, 87, 88
Epikur 57
Ernst, P. 114
Everett, H. 86
Feigl, H. x, xv-xviii, xxi-xxiv, xxvi, 5, 111, 113
Feyerabend, P. passim
Feynman, R. 83
Fischer, K.R. vii, 8, 77, 89, 92, 106, 113, 136
FitzGerald, G.F. 88
Flamm, D. 76
Floyd, J. vii
Foerster, H. v. xviii
Frank, Ph. x, xviii, xx, xxi, xxv, xxvii
Frege, G. 110, 111, 112, 114, 122, 129, 139
Frenkel-Brunswik, E. 1
Freud, S. 1, 89
Freundlich, R. xvii
Gabriel, L. 35
Galilei, G. xx, 43, 44, 45, 46, 58, 68-71, 78, 110
Gauss, C.F. 78
Geiger, T. xxi
Gellner, E. 101, 110
Giotto 111, 116
Gödel, K. 110, 111

Namenregister / Index of Names

Goldberger de Buda, R. xi
Gomperz, H. xvi
Grangier, Ph. 88
Grebovicz, M. ix
Grimm, J. u. W. 114
Grümm, H. xix
Grünbaum, A. xxv
Haberler, G. xviii, xxi
Hacker, P.M.S. 99
Hall, E. 78
Haller, R. vii, ix, 3, 4, 5
Hanisch, E. xviii
Hanson, N.R. xxiv
Hardy, G.H. 110
Hayek, F. v. x, xviii, xxi
Hegel, G.F.W. 4, 8, 18, 111
Heidegger, M. 4
Heintel, E. 4
Heisenberg, W. 19
Hempel, C.G. xvi, xxiv, 127
Hertz, H. 87
Hesse, M. xxiv
Hilbert, D. 110
Hintikka, J. xiv
Hitler, A. 8
Hjelmslev, L. xiv
Hochkeppel, W. ix
Hollitscher, W. x, xii, xix, xxviii, 105
Holton, G. xxviii
Hooke, R. 78
Horkheimer, M. xviii, xxi
Hoyningen-Huene, P. (P.H.-H.) vii, 13
Humboldt, A. 62
Hume, D. 9, 127
Huxley, A. 78
Jammer, M. 83
Jantsch, E. xi
Jedinger, M. 79
Joergensen, J. xii, xiv, xv
Juhos, B. x, xvii
Kaila, E. xiv
Kainz, F. xiv, 6
Kaller, R. viii
Kanitscheider, B. xxi
Kant, I. 42, 92, 111, 123
Karplus, R. 6, 77
Kaufmann, W. xxi
Kelsen, H. xxi
Kepler, J. 45, 46
Keupink, A. xvi
Knoll, R. vii
Köhler, W. 13, 15, 16
Kopal, Z. 47
Köstler, A. xviii

Kraft, V. x-xv, xvii, xx, xxii, xxvi, xxviii, 5, 35, 37, 76, 117, 118
Kraus, K. 132, 138, 140
Kreisky, B. xviii
Krenek, E. xviii
Kroß, M. 10
Kuhn, T.S. vii, 4, 15, 16, 19, 25, 35, 39, 46, 49, 51, 68, 69, 76, 79, 109, 126, 128
Lacan, J. 64
Lakatos, I. xviii, xxiv, 9, 46, 51, 76, 89, 107
Laplace, P.-S. 78
Larmor, J. 88
Leibniz, G.W. 9, 125
Lenin, W.I. xix
Leonardo da Vinci 55
Leukippos 57
Limbeck-Lilienau, C. viii
Locke, J. 9, 126
Lorentz, H. 88
Löwith, K. xxi
Mach, E. vii, xi, xvi, xxiv-xxix, 35, 36, 40, 45, 46, 72, 80, 108, 114
Machlup, F. xviii, xxi
Maier, A. xviii
Marcuse, H. xviii, xxi
Margolis, J. xxiv
Marx, K. xix
Mauer, O. 3
Maxwell, G. xvi, xxi, xxiii, xxiv
Meitner, L. xviii
Mill, J.St. 62, 63, 70
Mises, R. v. xxv
Molden, F. x, xviii
Molden, O. x, xviii
Monroe, M. 66
Moore, G.E. 6, 112, 138
Moser, S. xviii
Motterlini, M. xxiv
Mozart, W.A. 56
Muhammed Ali 66
Muhr, P. 8
Naess, A. xiv, xv
Nagel, E. xv, 19
Nestroy, J. 9, 56, 106, 130, 131, 132, 138, 140
Neumann, J. v. 77, 84, 119
Neurath, O. xiv, xv, xxv, xxvii
Newton, I. 45, 46, 78, 79
Nietzsche, F. 9
Noack, K.-P. xix
Oberheim, E. (E.O.) vii, 13, 16
Oeser, E. vii

Namenregister / Index of Names 155

Offenbach, J. 56
Pap, A. xv, xvi, xvii, xxi, xxiii
Parsons, T. 56
Peirce, C.S. 20
Petzäll, A. xiv
Pitcher, G. 122
Platon 2, 65
Podolsky, B. 81, 84
Popper, K. x-xii, xvi, xviii, xix, xxi, xxii, xxv, xxvii, xxviii, 9, 20, 23, 25, 35, 36, 40, 49, 51, 76, 77, 80, 81, 99, 101, 105-108, 110, 111, 114, 119
Przibram, K. x
Putnam, H. 127, 128, 129
Quine, W.V.O. xiv, 4, 99, 100, 102-104, 106, 111-113, 125-140
Radler, J. xi
Radner, M. xxiv
Reichenbach, H. xviii, 5, 54
Reisch, G. xx
Riegl, A. xxiv, 55
Roger, G. 88
Rohracher, H. xiv
Rorty, R. 99
Rosen, N. 81, 84
Rossi, P. 47
Rozeboom, W.W. xxiv
Russell, B. 6, 110-112, 114, 122, 129
Ryle, G. 121, 135
Sagan, J. x
Schelling, F. 4
Schimanovich, W. (Jimmy) 76
Schiske, P. xi
Schleichert, H. xvii
Schlick, M. xiv, xvii, xxi, 3
Schmid, E. xiv
Schrödinger, E. xviii, 83, 87
Schubert, F. 56
Schumann, R. 56
Sellars, W. xvii
Shakespeare, W. 121
Shie, S. xvi

Sieyès, E.J. 54
Sluga, H. vii
Snell, B. 65
Soames, S. 99
Sokal, A. 82
Specker, E. 84
Spinoza, B. 9
Stace, W.T. 83
Stadler, F. vii, xxv, xxviii
Stapp, H. 76
Stegmüller, N. x, xvii, xxi, 35
Stevenson, Ch. xv
Svozil, K. vii, 76
Tarski, A. 5, 113
Thirring, H. x, xi, xiv, xxviii, 6, 76
Tooke, J.H. 126
Topitsch, E. x, xvii, xxi
Tranekjaer-Rasmusssen, E. xi, xii, xiv
Urmson, J.O. 135, 137
Voegelin, E. 54
Voltaire 127
Wagner, R. 56
Waismann, F. 104
Walter, E.J. xi, xvii
Wegeler-Schardt, C. 8
Weiberg, A. 92
Weingartner, P. xxi
Wheeler, J.A. 83
Whewell, W. 35, 45, 46
Whitehead, A.N. 111
Wimmer, F. 8
Winokur, S. xxiv
Wittgenstein, L. vii, xi, xii, xvi, xxii, xxvii, 6, 9, 15, 19, 99-127, 129, 130, 131, 132, 133, 134, 136, 137, 138, 139, 140, 141
Wohlgenannt, R. xxi
Wolf, M. 47
Wolters, G. xxviii
Wright, G.H. v. xi
Zinner, E. 47
Zurek, W.H. 83

DIE AUTOREN / THE AUTHORS

Kurt Rudolf Fischer

Geboren 1922 in Wien. Flucht nach Brünn (1938) und Shanghai (1940). Studien an der St. John's University (Shanghai) und der Universität Wien. Abschluss an der University of California (Berkeley): M.A. (Germanistik, 1952), Ph.D. (Philosophie, 1964). Seit 1979 Honorarprofessor am Institut für Philosophie der Universität Wien; seit 1992 vom Gesundheitsministerium als Psychotherapeut approbiert. Zahlreiche Professuren und Dozenturen in Deutschland, Österreich und den USA, u.a. in Berkeley, Chicago und Harvard. Publikationen u.a.: *Nietzsche und das 20. Jahrhundert. Existentialismus. Nationalsozialismus. Psychoanalyse. Wiener Kreis.* (Wien 1986). *Philosophie aus Wien. Aufsätze zur Analytischen und österreichischen Philosophie. Zu den Weltanschauungen des Wiener Fin-de- Siècle und Biographisches aus Berkeley, Shanghai und Wien.* (Wien–Salzburg 1991). *Aufsätze zur angloamerikanischen und österreichischen Philosophie.* (Wien 1999). Herausgeber: *Österreichische Philosophie von Brentano bis Wittgenstein. Ein Lesebuch* (Wien 2000).

Juliet Floyd

Associate Professor of philosophy at Boston University, and a specialist on Kant, Wittgenstein, and the history of twentieth century – especially early analytic – philosophy. She has written articles on Kant, Frege, Wittgenstein, Gödel and Quine, as well as co-editing (with Sanford Shieh) *Future Pasts: The Analytic Tradition in Twentieth Century Philosophy* (Oxford University Press, 2001). She is currently completing a book on Wittgenstein's reactions to Gödel and Turing.

Paul Hoyningen-Huene

Geb. 1946. Studium der Physik und Philosophie in München, London und Zürich. Promotion in theoretischer Physik an der Universität Zürich 1975. An der Universität Zürich Assistent am Institut für Theoretische Physik 1972–1976, am Philosophischen Seminar 1975–1980; Lehraufträge für Philosophie an der Universität Bern 1980–1998. 1984–1985 Visiting Scholar am M.I.T., USA; 1987-1988 Senior Visiting Fellow am Center for Philosophy of Science, University of Pittsburgh, USA. Habilitation für Philosophie der Wissenschaften an der ETH Zürich 1988. 1989–90 Oberassistent für Umweltnaturwissenschaften an der ETH Zürich. 1990–1997 Professor für Wissenschaftsphilosophie und -geschichte an der Universität Konstanz; seit 1997 Professor für Ethik in den Wissenschaften und Leiter der Zentralen Einrichtung für Wissenschaftstheorie und Wissenschaftsethik an der Universität Hannover.

Reinhold Knoll

Geb. 1941 in Wien. Studium der Geschichte und Kunstgeschichte an der Universität Wien, Dr.phil. 1969 Verehelichung mit Dr. Elisabeth Biedl. Der Ehe entstammen zwei Kinder, Bernhard (1972) und Barbara (1975). Von 1970 bis 1972 Redakteur für Innenpolitik im Österreichischen Rundfunk/Hörfunk. 1973 Univ. Ass. am Institut für Soziologie der sozial- und wirtschaftswissenschaftl. Fakultät der Universität Wien, Habilitation für Geschichte und Theorien der Soziologie 1984, Univ.Doz., Mitglied der Arbeitsgruppe „Deutsche Geisteswissenschaften in den zwanziger Jahren" der Fritz-Thyssen-Stiftung. Seit 1994 Ass.Prof. mit der Abteilung für Kultursoziologie am Institut beauftragt, 1997 a.o.Univ.Prof. am Institut für Soziologie und Mitglied der Wiener Katholischen Akademie, 1998 Lehrbeauftragter an der reformierten Karoli-Gaspar-Universität in Budapest und seit 1998/99/00/01 regelmäßige Lehraufträge an der Universität Gödöllö für Kulturwissenschaften. Mitherausgeber gemeinsam mit M. Benedikt der Geschichte österreichischen Philosophierens: *Verdrängter Humanismus – Verzögerte Aufklärung.* Veröffentlichungen: *Zur Tradition der christlichsozialen Partei*, Wien 1973. *Österreichische Konsensdemokratie in Theorie und Praxis*, 1976. „Gründung des ‚Bundes der Arbeiterjugend Österreichs'"; in: *Dokumente, Etappen der katholisch-sozialen Bewegung in Österreich seit 1850*, St. Pölten 1980. „Der österreichische Beitrag zur Soziologie", in: *Soziologie in Deutschland und Österreich, Kölner Zeitschrift für Soziologie und Sozialpsychologie* 1981. *Protektion in Österreich*, Graz 1983. „Kognitive Dissonanz in Rechtsansprüchen", in: *Rechtstheorie, Recht als Sinn und Institution* 1984.

Die Autoren / The Authors

Eric Oberheim

Wissenschaftliche Hilfskraft am Lehrstuhl von Prof. Dr. Paul Hoyningen-Huene, Zentrale Einrichtung für Wissenschaftsgeschichte und Wissenschaftsethik, Universität Hannover. Forschung zu diversen Themen von Kuhns Einfluss auf die Soziologie bis zur Geschichte des Wiener Kreises. Bis Januar 2004 Promotionsstudium in Wissenschaftsphilosophie Zentrale Einrichtung für Wissenschaftstheorie und Wissenschaftsethik, Universität Hannover; Betreuer: Prof. Dr. Paul Hoyningen-Huene. Doktorarbeit: „On Feyerabend's Early Philosophy", abgeschlossen Mai 2004 mit *summa cum laude*.

Erhard Oeser

Geb. 1938 in Prag. Seit 1972 o. Prof. für Philosophie und Wissenschaftstheorie an der Universität Wien. 1986/1987 und ab 1994 Vorstand des Instituts für Wissenschaftstheorie und Wissenschaftsforschung der Universität Wien; seit 1984 Vorstandsmitglied der Österreichischen Gesellschaft für Geschichte der Wissenschaften. 1993 Gastprofessor an der Technischen Universität „Otto von Guericke" Magdeburg. Seit 1995 korrespondierendes Mitglied der Gesellschaft der Ärzte in Wien. Ab 1998 Vizepräsident und wissenschaftlicher Leiter des Karl Popper Institutes. Veröffentlichungen (u.a.): *Der erkenntnistheoretische Anarchismus* (1976), *Wissenschaftstheorie als Rekonstruktion der Wissenschaftsgeschichte* (1979), *Popper, der Wiener Kreis und die Folgen* (2003).

Hans Sluga

Hans Sluga is Professor of Philosophy at the University of California where he was for many years a colleague of Paul Feyerabend. He is the author of *Gottlob Frege* (1980) and *Heidegger's Crisis. Philosophy and Politics in Nazi Germany* (1993) and the editor of a four volume collection on *The Philosophy of Frege* (1993) and *The Cambridge Companion to Wittgenstein* (1998). He has written numerous articles on contemporary European philosophy and is currently finishing a book on political philosophy with the title *The Care of the Common*.

Die Autoren / The Authors

Friedrich Stadler

Studium der Geschichte, Philosophie und Psychologie in Graz und Salzburg. 1994 Habilitation für Wissenschaftsgeschichte und Wissenschaftstheorie an der Universität Wien. Seit 1997 außerordentlicher Professor an der Universität Wien, Zentrum für überfakultäre Forschung und ab 2001 Vorstand des Instituts für Zeitgeschichte. 1991 Gründer und seitdem Leiter des Instituts Wiener Kreis. Mitarbeiter des Ludwig-Boltzmann-Instituts für Geschichte und Gesellschaft. Zahlreiche Publikationen zur History and Philosophy of Science und zur Emigrationsforschung. Veröffentlichungen u.a.: *Vom Positivismus zur „Wissenschaftlichen Weltauffassung"* (1982); *Vertriebene Vernunft.* 2 Bde (1988, Neuauflage 2004); *Studien zum Wiener Kreis.*, 1997 (englische Übersetzung: *The Vienna Circle.* 2001); Hg.: *Elemente moderner Wissenschaftstheorie* (2000); gem. mit P. Weibel, *The Cultural Exodus from Austria* (1995); gem. mit M. Heidelberger, *Wissenschaftsphilosophie und Politik* (2003); *Induction and Deduction in the Sciences* (2004, = Vienna Circle Institute Yearbook 11/03).

Karl Svozil

Born 1956 in Vienna. Permanent position ("wissenschaftlicher Beamter") at the Institut für Theoretische Physik of the University of Technology Vienna. 1997 Assistenzprofessor at the Institut für Theoretische Physik of the University of Technology Vienna. 1997- A.o Univ. Professor at the Institut für Theoretische Physik of the University of Technology Vienna. More than one hundred publications in scientific research journals, two scientific monographs.

GPSR Compliance

The European Union's (EU) General Product Safety Regulation (GPSR) is a set of rules that requires consumer products to be safe and our obligations to ensure this.

If you have any concerns about our products, you can contact us on

ProductSafety@springernature.com

In case Publisher is established outside the EU, the EU authorized representative is:

Springer Nature Customer Service Center GmbH
Europaplatz 3
69115 Heidelberg, Germany

www.ingramcontent.com/pod-product-compliance
Lightning Source LLC
LaVergne TN
LVHW040738250326

834688LV00031B/350